U0424734

普通高等学校“十四五”规划英语实践实训数字化精品教材

英语同声打字教程

（第二版）

A Coursebook of English Simultaneous Typing

主　编：阮广红　龚一凡

副主编：夏胜武　刘　敏

编　者：裴　沁　鲁　萌　李莞婷

華中科技大學出版社

http://www.hustp.com

中国·武汉

图书在版编目(CIP)数据

英语同声打字教程 / 阮广红，龚一凡主编．— 2版．— 武汉 ：华中科技大学出版社，2021.8（2024. 8重印）
ISBN 978-7-5680-7478-0

Ⅰ．①英… Ⅱ．①阮… ②龚… Ⅲ．①商务-英文-文字处理-打字-教材 Ⅳ．①TP391.14

中国版本图书馆CIP数据核字(2021)第167142号

英语同声打字教程(第二版)　　阮广红，龚一凡　主编
Yingyu Tongsheng Dazi Jiaocheng(Di-er Ban)

策划编辑：周晓方 宋　焱
责任编辑：周清涛
封面设计：原色设计
责任校对：张汇娟
责任监印：周治超
出版发行：华中科技大学出版社(中国·武汉)　　电话：(027)81321913
武汉市东湖新技术开发区华工科技园　　邮编：430223
录　　排：湖北新华印务有限公司
印　　刷：武汉开心印印刷有限公司
开　　本：787mm×1092mm　1/16
印　　张：10.25　　插页：2
字　　数：300千字
版　　次：2024年8月第2版第2次印刷
定　　价：58.00元

前言

Preface

“英语同声打字”是一门英语实训课程，旨在通过同声听打英语练习，使学生的英文打字速度达到一定的水平，提高学生计算机的实操能力，让学生在练习的过程中提高英语听力水平，巩固英语各个方面的知识，包括商务信函、商务沟通、贸易流程、媒体与宣传等方面的知识，帮助学生夯实语言基础，拓展人文百科知识，提高英语综合应用能力。本课程的实践练习分为三个阶段：键盘输入、英文听打和模拟现场英语同声打字。通过大量的英语同声打字实践，学生能将听到的所有内容或主要内容以大纲、笔记的形式在规定的时间内输入电脑，形成可读性强且忠于听力材料的英文文章。

新版教材按照实际的贸易流程对老版教程进行了结构优化，更有利于学生商务知识体系的整体性建构。本教材的主要特点有：一、所选内容兼具时效性和科学性，强调选材的时代性、多样性和可行性，尽量选用最新的英语语料，涵盖商务英语学习的主要板块和当前的热点话题；在结构编排上从易到难，避免单一类型材料贯彻始终，做到难度适中，因材施教。二、教材设计兼具人文性和技术性，兼顾学生英语综合运用能力（尤其是英语听力）和计算机操作能力的培养。三、思政融入。教学内容和练习设计精选思政素材，将思想政治教育内化为实训教材内容，将价值塑造、知识传授和能力培养融为一体。

本教材由五个部分组成：第一部分介绍打字基础，包括打字的基本要领及指法，同声打字的基本术语及信函的基本格式和同声打字常用缩写；第二部分为商务信函，指导学生听打英语便函、电话留言、会议纪要和报告；第三部分为商务沟通，涉及邀请与建立业务关系、询价与回复、报盘与还盘、订单等内容；第四部分介绍贸易流程，包括商务合同、支付、发货和售后服务；第五部分为媒体与宣传，包括礼仪祝词、新闻发布会、广告宣传、商务新闻和今日中国等。

本教材所选材料包含书信、对话、电话记录、便函、报告、合同等各类体裁。每章由预习和听打训练两个部分组成：预习部分包括生词练习、句子练习、段落练习、短文或对话练习；听打训练部分包括简单句子练习、复杂句子练习、段落练习、短文或对话练习和注释。该门课程为32学时，学生可以在教师的指导下依次按章节进行听打训练，也可自主进行听打训练。标准听打英语的速度应该在每分钟

120个字符左右。

本教材具体编写分工为:刘敏第1章第1、2、3部分;李莞婷第2章第1、4部分;夏胜武第2章第2、3部分;裴沁第3章第1、2、3、4部分;龚一凡第4章第1、4部分和第5章第5部分;鲁萌第4章第2、3部分;阮广红第5章第1、2、3、4部分。曹曼负责全书的主审工作。

本教材具有较强的针对性和实用性,并配有音频材料,适合商务英语、英语和翻译专业的综合实训课程教学,也适于具有一定英语基础的学习者自主训练。在教材编写过程中,我们得到了华中科技大学出版社的大力支持,在此深表感谢。由于时间和水平有限,书中难免存在错漏之处,敬请专家和读者们批评指正。

编　者

2021年4月

目录

Chapter 1 Typing Elements

打字基础

1.1 The Main Points and Fingering 打字的基本要领及指法

Part One The Formation and Function of a Keyboard 键盘的构成与功能

计算机键盘是用户向计算机输入信息，控制计算机操作的主要输入设备，它主要由五个部分组成：功能键区、主键盘区（也叫打字键区）、编辑键区、状态指示区和辅助键区（也叫数字键区）。功能键区（共13个）在键盘的第一排，包括Esc键和F1—F12键。主键盘区（打字键区，共61个）包括字母键A—Z共26个，数字键0—9共10个，符号键和其他一些功能键共25个。编辑键区（共13个）包括上（↑）、下（↓）、左（←）、右（→）4个方向键及其上方的9个键。辅助键区（数字键区，共17个）包括0—9共10个数字键及其他7个功能键。

键盘布局如图1-1所示。

图1-1 键盘布局

1. The Functional Keys on the Typing Keypad 主键盘区主要功能键

Back Space	后退键,删除光标前面的字符。
Enter	回车键,亦可叫做换行键,将光标移至下一行的行首。
Shift	换档键,与字母键同时按下时可打出大写字母,与数字键或符号键同时按下时可打出该键的上部符号。
Ctrl	控制键,必须与其他键配合使用。
Alt	选择功能键,必须与其他键配合使用。
Tab	跳格键,将光标右移到下一个跳格位置。
Caps Lock	锁定键,将英文字母锁定为大写状态。
Space Bar	空格键,输入空格。

功能键区F1—F12的功能根据具体操作系统或应用程序而定,在编辑Word文档时一般不会使用它们。

2. The Functional Keys on the Editing Keypad 编辑键区主要功能键

Print Screen / SysRq	复制当前屏幕上所显示的所有内容。
Scroll Lock	屏幕锁定键,按下此键屏幕停止滚动,直到再次按此键为止。
Pause Break	暂停键,按下此键,暂停系统运行(屏幕停止滚动)。
Insert	插入改写键,按下此键,输入文字时将删除光标后的字符。
Home	此键可将光标移至行首,配合**Ctrl**键使用时,可将光标移至文档开头。
Page Up	上翻页键,按下此键光标移至上一页。
Delete	按下此键将删除光标后字符。
End	按下此键,光标将移至行尾,配合**Ctrl**使用时,可将光标移至文档末尾。
Page Down	下翻页键,按下此键时,光标将移至下一页。

3. The Shortcut Keys and Their Functions Used in Word 编辑Word文档时常用的快捷键及其功能

Ctrl + A	将两键同时按下,可选中当前编辑文档的所有内容。
Ctrl + B	将两键同时按下,可将选中的内容变为粗体。
Ctrl + I	将两键同时按下,可将选中的内容变为斜体。
Ctrl + C	将两键同时按下,可将选中的内容复制到剪贴板中。
Ctrl + X	将两键同时按下,可将选中的内容剪切到剪贴板中。
Ctrl + V	将两键同时按下,可将剪贴板中的内容复制到光标处。
Ctrl + F	将两键同时按下,可打开“查找和替换”对话框。
Ctrl + N	将两键同时按下,可建立一个新文档。

Ctrl + P	将两键同时按下,可打开"打印"对话框。
Ctrl + S	将两键同时按下,可对当前编辑文档进行保存。
Ctrl + Z	将两键同时按下,可撤销上次操作。
Ctrl + Shift	将两键同时按下,可对输入法进行转换。
Ctrl + Enter	将两键同时按下,可将光标移至文档下一页。
Shift +字母键	将两键同时按下,可输入大写字母。
Alt +F4	将两键同时按下,可关闭当前活动窗口。

Part Two　English Touch System 英文打字指法

打字时,除了要对键盘特别熟悉外,还必须掌握正确的打字指法。只有掌握了正确的指法,打字速度才能提高。

打字时,每个手指都有各自的分工,都分配有基本键。除拇指外,其余的8个手指分别放在基本键上,拇指放在空格键上,十指分工,包键到指,分工明确。

、,1,Tab,Caps Lock,Shift,Ctrl,Alt,Q,A,Z键位由左手小指负责;2,W,S,X键位由左手无名指负责;3,E,D,C键由左手中指负责;4,R,F,V,5,T,G,B键由左手食指负责;6,Y,H,N,7,U,J,M键由右手食指负责;8,I,K,<键由右手中指负责;9,O,L,>键由右手无名指负责;空格键由大拇指负责;其余的键由右手小指负责。具体如图1-2所示。

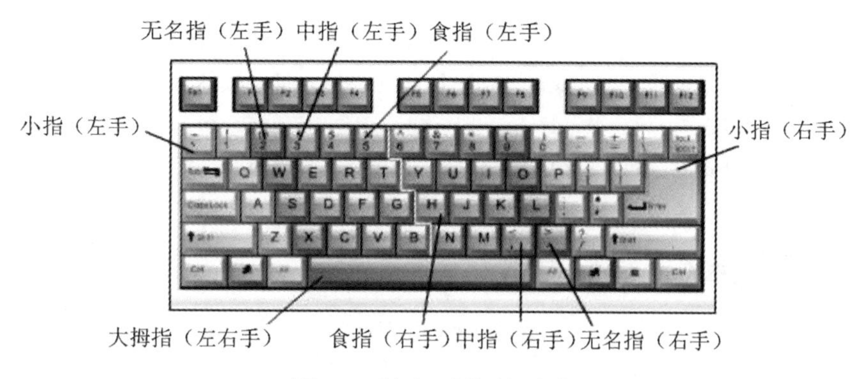

图1-2　键与手指的对应

注意:准备打字时,手指要轻轻地放在键盘的基本键上(ASDFJKL;),左手食指放在F键上,左手中指、无名指、小指分别放在D,S,A键上,右手食指放在J键上,右手中指、无名指、小指分别放在K,L,;键上。在敲击了其他各键后,手指应该迅速回到原来键位上。在停止打字的状态下,手指要始终保持在基本键位上。

Part Three Typing Posture 打字坐姿

打字时,一定要有正确的坐姿,如果坐姿不正确,不但容易疲劳,而且还很容易出错。

正确的打字姿势为:

① 两脚平放在地上,腰部挺直,两臂自然下垂,两肘贴于腋边;

② 身体可略倾斜,离键盘的距离约为20~30 cm;

③ 打字的材料或文稿放在键盘的左边,打字时眼观文稿,身体不要跟着倾斜;

④ 同声打字时,思想一定要高度集中。

Part Four Basic Fingering Practice 基本指法练习

练习1 字母练习

a s d f j k l ;

asdf	asdj	asdk	asdk	sadf	sadj	sadk
sadl	dasf	dasj	dask	dasl	fdad	fasd
fdad	fjkl	fjka	fjks	fjkd	fjkj	aass
ss;dd	d;dff	ffjj	jjk;k	k;kll	kl;d	df;j

练习2 字母练习

a s d f g h j k l ;

asdfg	asdfh	asdfk	asdfl	sdfga	sdfha	sdfja
dsafg	dsfgkj	sdfgkl	ghjkl	sdfgh	dfghj	fjghk
lkjh	ljhg	lasdf	kasdf	glkas	ghjkd	jkefg
jdsfl	lghjs	sjkgh	dfjkh	hfksl	ghj;;g	as;g;k

练习3 字母练习

q w e r t y u i o p

qwertt	qwerty	qeruip	wuriep	rutieo	qieuru	yitoew
urueoi	reieoo	ruiow	qpeori	ruieow	ieiowo	tuieow
qowue	rueiwe	rptoru	rtueiw	qoeiru	turyei	turieo
puoyit	iutyre	uywtq	riotorp	wueur	tuyyeo	riwoqp

练习4 字母练习

z x c v b n m

zxcvb	zcvbn	zxcvn	zvnbn	zmxnc	vbcnx	cmzcb
zcvnm	mnbcv	mxbcv	zxmnb	cvmnx	xcmnz	cvbnx

vncxm	zxcvb	vbcnx	mnxcv	bvnxz	nbxmv	nbxcv
mznxc	nxcbz	bvnmx	xcbvz	vbncx	ncmxb	xmvnz

练习5 字母练习

A—Z 键及符号键

qdweu	wsjdeu	fyurhr	njuyh	nkilow	khgre	vcdwy
hytrdb	vdetyu	wsfgtr	jhytfv	knyrdc	iklotr	vryjjbe
vfrhkj	rdewsh	vgtrhu	njktrf	wsgrfj	bvrgty	cfujmn
‘ ; ./	, ; [‘	[ ] ‘ ‘	; /. ‘	[; ’ [	[/ ‘ ;	
> <?”	{ } : ”	? > { :	( ) _+	} { ? >	{ [? ∧	

练习6 单词练习

jovial	behalf	nearby	pigsty	fabled	yellow
purely	should	sandal	blazed	labour	thresh
follow	jasper	cattle	colour	fifth	intend
future	evolve	hourly	weekly	monthly	mother
little	rundle	distal	around	Easter	useful

练习7 单词练习

north	zero	essay	next	giant	handle
house	tramp	three	rough	sexy	eleven
height	brisk	annul	civil	south	temp
differ	golden	rosy	cubbish	quest	wives
often	exam	error	chief	night	Kodak

练习8 单词练习

tortuous	manufacture	housewife	hospital	handmade	magnificent
maroons	salable	embroidered	catalogue	business	insurance
reasonable	possibility	importance	anticipate	meantime	presently
equipment	workman	information	including	exporters	handbags
producers	equipment	underground	yesterday	shipment	colouring

练习9 句子练习

1. I am always suspicious of anyone who wants to sell me something on the cheap.
2. He is the only successor among his brothers of their father.
3. Boeing 747 plane is a huge, luxurious supersonic passengers' plane.
4. Dolphins can be useful for the submarine explorations if trained properly.
5. They still remembered each other after a stretch of twenty years.
6. We have striven to the full to convince him, but we have made no headway.
7. Susan likes to stroll around the flea market on the chance of picking up something of value.
8. Sociology deals with the facts of crime, poverty, marriage, divorce, the church, the school, etc.
9. The perennial conflict between national egoism and international solidarity becomes more and more visible.
10. Many scientists remain skeptical about the value of this research program.
11. The liquid is purified by passing it through charcoal.
12. What are the prospective returns from an investment of 20,000 over three years?
13. The delegation decided to prolong their visit by five days.
14. Although this area is very poor now, its potential wealth is great.
15. The X-ray is a practicable way of discovering diseases.
16. The tourists were preceded by their guide.
17. She preceded her speech with welcome to the guests.
18. Mike should get a lawyer to plead his case.
19. The principles of the two methods are completely polar.
20. Applying for a job now starts with rewriting and updating your CV.

练习10 句子练习

1. The mistakes of the ministers provided perfect ammunition for their political enemies.
2. Let me amplify so that you will understand the overall problem.
3. There is an analogy between the way water moves in waves and the way light travels.
4. A man who always anticipates his income can never save or become rich.
5. The residents in the neighborhood all applauded the council's decision to close the small dye factory.
6. This encyclopedia has a supplement covering recent events.

7. China's number of online short-video users reached 873 million in 2020.
8. John is so arrogant that he thinks he is better than everybody else.
9. He gave me a definite assurance that the repairs would be finished tomorrow.
10. Is that an authentic painting from Picasso, or a modern copy?
11. The harvest is better on an average this year.
12. We tried and tried, but it was all to no avail: we failed.
13. A summer at the seashore benefits the entire family.
14. The earth rotates about an axis between the North Pole and the South Pole.
15. To do the job, you must have at least a bachelor degree in science.
16. The old woman from the country was bewildered by the crowds and traffic in the big city.
17. They prepared a special dinner which was so elaborate as to become a banquet.
18. It is not the quantity of the food, but the cheerfulness of the guest, which makes the feast.
19. The remote barren land has blossomed into rich granaries.
20. These are the last batch of letters to be answered.

练习 11 句子练习

1. The crowd's chant was "More Jobs! More Money".
2. We never cherish any unrealistic fancies about those desperate criminals.
3. He carried a board, onto which his secretary had clipped all the important documents.
4. The clearance between the bridge and the top of the car was only ten feet.
5. His election to the presidency was the climax of his career.
6. I am determined to face the challenge whatever may come to me.
7. The building was built to commemorate the Fire of London.
8. One should refrain from applause during a debate.
9. I argued with her for a long time, but she refused to listen to reason.
10. He and his wife had cheated every one with whom they had dealings.
11. An accident has disrupted railway services into and out of the city.
12. The noise of cars passing along the road is a continual disturbance to our quiet at home.
13. I think the present armed clash on the border was a diversion to make their people forget the internal difficult economic situation.
14. The company said the rules should be put in place next year.
15. Finding high-quality candidates who have the right skills is not an easy task.
16. The newspaper extracted several passages from the speech and printed them on the

front page.
17. These superstitious practices should be abolished as soon as possible.
18. He has flushed with excitement when he learned that he had won the first prize.
19. The treaty will give an impetus to trade between Russia and China.
20. The Export Commodities Fair was inaugurated the day before yesterday.

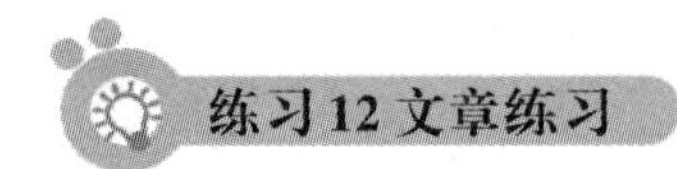

练习 12 文章练习

Yellow Crane Tower

Yellow Crane Tower is an imposing tower close to the Changjiang River. Situated at the top of Sheshan Hill(Snake Hill), in Wuchang, the tower was originally built at a place called Yellow Crane Rock projecting over the water, hence the name. Over the centurics the tower was destroyed by fire many times, but its popularity with Wuhan residents ensured that it was always rebuilt. The current tower was completed in 1985 and its design was copied from a Qing Dynasty picture. The tower has 5 stories and rises to 51 meters covered with yellow glazed tiles and is supported with 72 huge pillars. It has 60 upturned eaves layer upon layer. It is an authentic reproduction of both the exterior and interior design, with the exception of the addition of air-conditioning and an elevator.

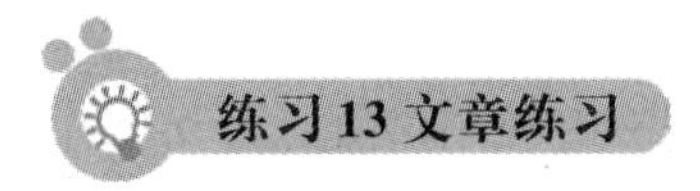

练习 13 文章练习

British Pub Culture

Visitors to Britain may find the best place to sample local culture is in a traditional pub. But these friendly hostelries can be minefields of potential gaffes for the uninitiated.

An anthropologist and a team of researchers have unveiled some of the arcane rituals of British pubs—starting with the difficulty of getting a drink. Most pubs have no waiters—you have to go to the bar to buy drinks. A group of Italian youths waiting 45 minutes before they realized they would have to fetch their own. This may sound inconvenient, but there is a hidden purpose.

Pub culture is designed to promote sociability in a society known for its reserve. Standing at the bar for service allows you to chat with others waiting to be served. The bar counter is possibly the only site in the British Isles in which friendly conversation with strangers is considered entirely appropriate and really quite normal behaviour. "If you haven't been to a pub, you haven't been to Britain." This tip can be found in a booklet, Passport to the Pub: The Tourists' Guide to Pub Etiquette, a customers' code of conduct for those wanting to

sample “a central part of British life and culture”. The trouble is that if you do not follow the local rules, the experience may fall flat. For example, if you are in a big group, it is best if only one or two people go to buy the drinks. Nothing irritates the regular customers and bar staff more than a gang of strangers blocking all access to the bar while they chat and dither about what to order.

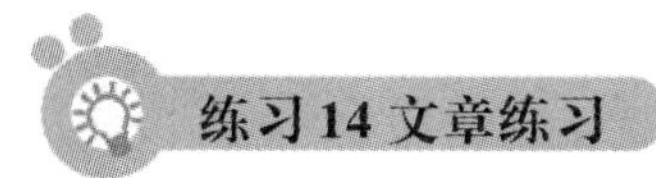

Goal for 2023: 560 Million 5G Users in Nation

China aims to nurture 560 million 5G mobile subscribers by the end of 2023, and grow the penetration rate of the fast wireless technology among big industrial enterprises to 35 percent by then, a three-year plan jointly unveiled by 10 government bodies disclosed on Tuesday.

The announcement of the twin goals reflects China's vigorous promotion of the comprehensive and coordinated development of 5G and its efforts to further widen the use of 5G in order to empower a wide range of industries.

The three-year plan was published by the Ministry of Industry and Information Technology and nine other government bodies. The application of 5G in different sectors, it said, is important to promote the digital, networked, and intelligent transformation of the economy and society.

The number of 5G connections has exceeded 365 million, accounting for 80 percent of the world's total. According to the three-year plan, China aims to grow the penetration rate of 5G among individual consumers to 40 percent by the end of 2023, with 5G data accounting for more than half of overall online traffic.

In the enterprise market, the plan called for efforts to popularize the use of 5G in big industrial enterprises, and to scale up its application in grid, mining and other sectors. The scope of pilot projects involving 5G plus connected vehicles will be further expanded and efforts will also be made to accelerate the digital transformation of the agriculture sector.

To lay sound telecom infrastructure, China will accelerate the rollout of 5G networks, with the aim of having more than 18 5G base stations per 10,000 people by the end of 2023.

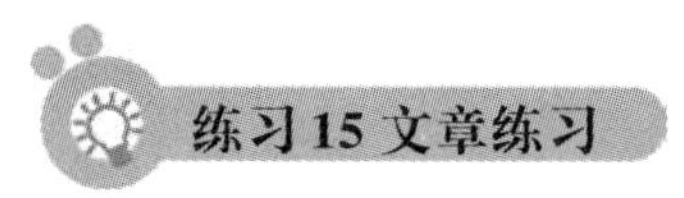

China National Import & Export Corp.
Shenzhen Branch
Shenzhen, China

15th March, 2021

Roman International Inc.
659 Peace Road
Alabama
USA

Dear Sir / Madam,
Thank you for your letter of 15th August, enquiring for Happiness Brand Bassinet.
We are exporting bassinets of various brands among which Happiness Brand and Hong Yun Brand are the most famous ones. They are in great demand abroad and our stocks are running down quickly. They are popular not only for their reliable qualities, but also for the reasonable prices. We are confident that once you have seen our bassinets you will place repeat orders with us in large quantities.
Based on your requirement, we are quoting as follows:
Happiness Brand Bassinets:
Luxurious $58 / piece
Economical $48 / piece
Hong Yun Brand Bassinets:
Luxurious $52 / piece
Economical $42 / piece

A discount of 5% may be allowed if the quantity for each specification is more than 150 sets.
The above quotation is made without engagement and is subject to our final confirmation.
We look forward to your early reply.
Yours faithfully,

1.2 Basic Terms and Letter Formats 基本术语及信函的基本格式

Part One Basic Terms 基本术语

heading	信头
body	正文
paragraph	段落
bold	粗体
italic	斜体

	initial capitals	首字母大写
	indented form	缩进式
	block form	齐头式
	hanging-indentation form	悬挂式
.	period	句号
,	comma	逗号
:	colon	冒号
;	semicolon	分号
!	exclamation mark	感叹号
?	question mark	问号
-	hyphen	连字符
—	dash	破折号
__	underscored / underlined	下划线
’	apostrophe	撇号
‘ ’	single quotation marks	单引号
“ ”	double quotation marks	双引号
( )	parentheses	圆括号
[ ]	square brackets	方括号
...	ellipsis	省略号
‖	parallel	双线号
/	virgule	斜线号
&	ampersand = and	与
%	per cent	百分号
‰	per mill	千分号
°	degree	度
℃	Celsius system	摄氏度
′	minute	分
″	second	秒

Part Two Letter Formats 信函的基本格式

1. Inside Address 信内地址

信内地址包括发信人的地址和收信人的地址两部分。发信人的地址写在信纸的右上角；收信人的地址在发信人地址下一行的左边。在商务信函中，地址的内容一般包括姓名（单位名称）、地址、电话、传真、E-mail和日期等。英文地址和中文地址不同，要按照从小到大的顺序写；英文日期的格式与中文的格式也有所不同，如2021年7月15日，在英文中可写为:"July 15,2021"(较为普遍),"July 15th,2021","15th July,2021"等。

2. Salutation 称呼

写信人对收信人的称呼,位置在信内地址下方一、二行的地方,从该行的顶格写起,在称呼后面用逗号。

A. 写给亲人、亲戚和关系密切的朋友时,用Dear或My dear再加上表示亲属关系的称呼,或直称其名(这里指名字,不是姓氏)。例如:My dear sister,Dear Anna等。

B. 公务上的信函用Dear Madam,Dear Sir。

C. 写给收信人的信,也可用头衔、职位、职称、学位等再加姓氏或姓氏和名字。例如:Dear Prof. Green,Dear Dr. John Smith。

3. Body of the Letter 正文

一般在称呼语后换行开始写正文,其格式为缩进式(indented form)、齐头式(block form)和悬挂式(hanging-indentation form)三种。

缩进式 正文每段第一行的第一个字母向右缩进四个字符再开始写,其他行从左顶格写起;

齐头式 正文每段第一行都从左顶格写起;

悬挂式 正文每段第一行从左顶格写起,其他行从左向右缩进四个字符后再开始写。

商务信件大多采用齐头式的写法。

4. Complimentary Close 结束语

在正文下面的一、二行处,齐头书写结束语,亦可从中间偏右处开始书写,第一个词开头要大写,句末用逗号。

5. Enclosures 附件

商务信函中如果有附件,可在信纸的左下角,注上Encl:或Enc:,例如:Encl: A Brief Introduction of Shangri-La Hotel。如果附件不止一项,可写成"Encls:"或"Encs:"。

Part Three Sample Letters 信函格式样本

1. Indented Form 缩进式

Dear Ms. Roberts,

I noticed the advertisement in last Friday's newspaper. Your company seems to have many kinds of wrapper. Our company deals with souvenir. If the quality of your product comes up to our expectations, we may be able to place large orders with your company.

We would appreciate if you send us your latest catalogue and price list.

Yours sincerely,

Julie

Dear Mr. Gray,

Your catalogue and price list on July 26th was received. We feel pleased that you offered the terms we need. We are pleased to enclose an order form, No. NA2673.

Please let us informed if you plan to deliver them before the arrival date specified.

If our business, with your cooperation, goes as well as we hope, we shall place further order with you.

Yours sincerely,

Sarah

2. Block Form 齐头式

Dear Mr. Dupant,

The bearer of this letter, Ms. Smith, the manager of the Product Department of this corporation, is going to be in New York on Sept. 5th. She will be living in your hotel for three days.

I shall be very glad if you can assist Ms. Smith in anyway. She behaves both honestly and responsibly.

Yours sincerely,

James

Dear Mr. Lee,

I wrote this resignation with a mixed feeling. Because of the confidence you have shown in my ability and personality, my six months in this school have been a period with much growth and challenge.

But as you probably know, I am also very interested in English teaching. Now that I got the opportunity, I would like to have a try.

Thank you sincerely for all the support and encouragement you have provided during the months we worked together.

Yours sincerely,

Anna

3. Hanging-Indentation Form 悬挂式

Dear Mr. Williams,

Your 3,000 fans under our order No. NA748 were received on May 19th.

However, we regret to tell you that the quality of those fans is not good enough. Previously,

the quality of your goods is always quite reliable. But we are disappointed this time.
We have sent five of them to you. Please test them and write back soon.

Yours sincerely,

Harry

Dear Mr./Ms.,

Thank you for your letter of July 5th, enclosing an account of the organization and work of your Chamber of Commerce and Industry.

We are very grateful for such a detailed account of your activities. This information is certain to help increase our future cooperation.

Yours faithfully,

1.3 Common Abbreviations 同声打字常用缩写

英语同声打字过程中,常通过缩写单词或词组的形式来提高同声打字效率。

1. 去掉所有元音字母

prdct	product
mrkt	market
rspnsblty	responsibility
ngttn	negotiation
xpndg	expanding
rqrmnt	requirement
pymnt	payment
shpmnt	shipment
mnfctrr	manufacturer
rltnshp	relationship

2. 保留几个字母

info	information
dept	department
intnl	international
biz	business
com	company
uni	university

qnaire	questionnaire
pls	please
pl	people

3. 根据主题缩写词组（以 **establishing business relationship** 主题为例）

BR	business relationship
t/p	terms of payment
JV	joint venture
I&E	import and export
MB	mutual benefit
SM	Sales Manager
PL	price list
p-smt	prompt shipment
l-trm c /cprtn	long term cooperation
BN	business negotiation

4. 句子和段落缩写范例

1） We are willing to enter into a business relationship with your company on the basis of equality and mutual benefit.（96个字符）

W'r wling to entr into a BR wth ur com. on the bss of eqlty and MB.（51个字符）

2） We have never had any difficulties with our Chinese partners，and we'd like to make as many new contacts as we can.（94个字符）

W'v nvr hd any dfclties wth our C. prtnrs，and w'd lk to mk as mny nw cntcts as we cn.（66个字符）

3） If you are interested in dealing with us in other products of our company，please inform us of your requirements as well as your banker's name and address.（128个字符）

If u r intrstd in dlg wth us in othr prdcts of our com，pls infrm us of ur rqrmnts aw aur bnkr's nm & add.（81个字符）

4） With a view to expanding our business at your end，we are writing to you in the hope that we can open our business relations with your firm.（113个字符）

Wth a vw to expndg our biz at ur end，w'r wrtg to u in the hp tht we can opn our BR wth ur frm.（70个字符）

5） We have a long experience in the import and export trade and a wide knowledge of commodities as well as of the best sources of supply of these materials.（127个字符）

W'v a l-exprnc in the i&e trd and a wd knwldg of cmmdts awa of the bst srcs of

spply of ths mtrls.（76个字符）

6） In order to give you a general idea of various kinds of tablecloth we are handling，we are airmailing you under separate cover our latest catalogue for your reference. Please let us know immediately if you are interested in our products. We will send you our price list and sample to you as soon as we receive your specific enquiry.（275个字符）

In ordr to gv u a gnrl idea/vrs knds / tblclth w'r hndlg，w'r armlg u undr sprt cvr our ltst ctlg 4 ur ref. Pls lt us knw immdtly if u'r intrstd in our prdcts. W'l snd u our PL & smpl to u asa we recv ur spcfc enqry.（166个字符）

7） We are one of the largest computer manufacturers in our country and have handled various kinds of products for about 10 years. We approach you today in the hopes of establishing business relations with you and expect，by our joint efforts，to enlarge our business scope.（227个字符）

W'r 1/ the lgst cmptr mnfctrrs in our cntry and hv hndld vrs kds/p 4 abt 10yrs. We apprch u tdy in the hps / estblshg BR wth u and expct，by our j-effrts，to enlrg our biz scp.（138个字符）

8） We are exporting bicycles of various brands among which Forever Brand and Phoenix Brand are the most famous. They are in great demand abroad and our stocks are running down quickly. They are popular not only for their light weight，but also for their reasonable prices. We are confident that once you have tried our bikes you will place repeat orders with us for large quantities.（317个字符）

W'r exprtg bk / vrs brnds amng which FB and PHB r the mst fms. T'r in grt dmnd abrd and our stcks r rnng dwn qckly. T'r pplr not only for thr l-wght，but als for thr rsnabl prcs. W'r cnfdnt tht once u hv trd our bks u'l plc RO wth us 4 lrg qntties.（192个字符）

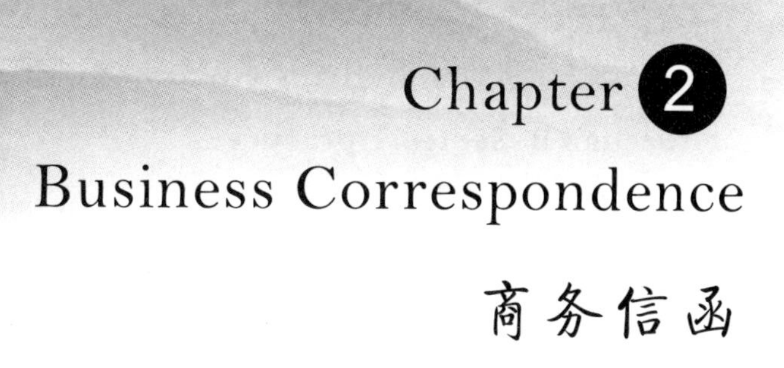

Chapter 2 Business Correspondence

商务信函

2.1 Memos 便函

Part One: Warm-up Activities

In Part One, you will practice typing sentences, paragraphs and passages dealing with memos. Firstly you are supposed to read aloud and to identify the new words listed in Section A, and then try to practice typing the sentences, paragraphs and passages in Section B, C, D respectively under the teacher's guidance. While you are typing, please mark out the time you spend on each section and compare your results with your classmates.

Section A Vocabulary work

memorandum	*n.*	便函;备忘录
CEO (chief executive officer)		首席执行官
convey	*v.*	表达;传递
professional	*a.*	专业的
recipient	*n.*	接收者
salutation	*n.*	(信函中如dear sir之类的)称呼语
format	*n.*	格式
attachment	*n.*	附件
appropriately	*ad.*	适当地
concise	*a.*	简明的
tailor	*v.*	专门定制

launch	*v.*	发布
extensive	*a.*	大量的;广阔的
promotional	*a.*	促销的
communal	*a.*	共有的
facility	*n.*	设备

Section B Sentence practice

1. Memorandum is a common form of communication within a company, usually called a memo.
2. Memos are written by everyone from junior executives and engineers to CEOs to convey information within an organization.
3. Understanding the correct memo format can help you communicate information more efficiently and professionally.
4. It is vital to use the correct format when you are writing a memo to ensure efficient communication.
5. You can use memos to ask for information, request confirmation or suggest an action.
6. You can also send a memo to provide a progress report to colleagues and other people in your organization.
7. All memos follow the same format, which has elements addressing the recipient, the sender, the date and the body.
8. A memo does not require salutation and may not include a sign-off.
9. You can send a memo either in printed form or as an email.
10. Start with a direct and brief introduction that states the reason for writing the memo.

Section C Paragraph practice

1. The heading segment of memo follows this general format:

   To:readers' names and job titles

   From:writer's name and job title

   Date:complete and current date

   Subject:what the memo is about, highlighted in some way

   Generally, you should clearly state your purpose at the beginning of the memo and request the action you want at the end. Address the recipient appropriately. A memo is a formal business communication, and you should address the reader formally as well. Use a full name and title of the person to whom you are sending the memo. If you are sending a memo to the entire staff, you might write: "TO: All Employees".
2. A memo is simply a way to communicate information within an organization. You can

use it to communicate with your employees, colleagues, or supervisors. It can be addressed to one recipient or to multiple people at once. A business memo should include a heading (which contains the to and from information), a date, a subject line, and the actual message of the memo. The body of the memo might contain an introduction, details that expand on the topic of the memo, and a request for some type of action from the recipients. You could also include attachments if necessary.

3. Write the complete date, spelling out the month and including the date and year. Choose a specific phrase for the subject line. The subject line gives the reader an idea of what the memo is about. Be specific but concise. Consider who the audience should be. In order to get people to read and respond to the memo, it's important to tailor the tone, length, and level of formality of the memo to the audience who will be reading it. Doing this effectively requires that you have a good idea of whom the memo is intended for.

Section D Passage practice

➢ Passage 1

MEMORANDUM

Date: October 13, 2020
To: All Staff
From: John Smith, Director, Product Development
Subject: Launch of Product Model R75

Due to extensive customer feedback, and the results of current testing, I want to inform you that Product Model R75 will be delayed from its original launch date of November 8th, until Q1 2021.

We are confident that time for additional testing will serve to make our R75 more effective in fighting security breaches that customers are facing. For those customers that you believe will now consider a competitor's product, the Marketing Department is developing a promotional offering, which sales reps can share with their customers to help reduce those who will now go to our competitors.

As disappointing as this news may be, we are confident in our employees, and know the additional time will serve this company well by creating a more successful product.

➢ Passage 2

MEMORANDUM

Date: November 17, 2020

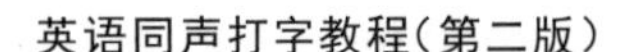

To: All Employees

From: Emma Johnson, VP, Marketing Department

Subject: Facilities Update

I'm writing to inform you that, over the next few weeks, our kitchen area will be under construction as we re-model it.

As our company continues to grow, we feel it necessary to provide more communal dining space, and we are grateful to our facilities team for their hard work in making that happen.

If you have questions or concerns regarding the re-model, you can access the full report here. In the meantime, we are sorry for the inconvenience.

In between the weeks of December 1 to December 31, please use the kitchen on the second floor if you need a microwave. We will also keep that kitchen stocked as per usual with snacks and soda.

Thank you for your cooperation.

Part Two: Audio-typing

Now you are going to listen to the recording. Do not refer to your textbook while you are listening. Then you are supposed to listen to each section sentence by sentence once again and type what you hear at the same time.

Section A

clip 2-1

Listen to the audio clip 2-1 and type the following short and simple sentences.

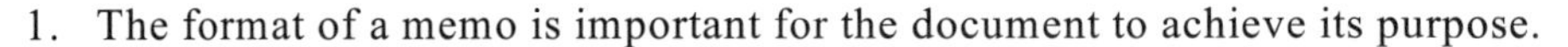

1. The format of a memo is important for the document to achieve its purpose.
2. If you use the wrong format, the memo may not deliver your message correctly.
3. Provide concise but detailed information to the reader.
4. A memo is one of the most important forms of communication used in public and private organizations.
5. All employees must use the new accounting system by June 1, 2021.
6. As of May 1, 2021, ABC Corporation will be implementing new policies regarding health coverage.
7. All employees will receive health coverage and will make a minimum of $15 per hour.
8. The county government voted to require all employees to receive a $15/hour minimum wage.
9. The Board urgently requires feedback on our experience with laptops in Eastern

Branch.

10. More information about this will be forthcoming from marketing.
11. Memos are a great way to communicate big decisions or policy changes to your employees or colleagues.
12. Staff are complaining about the poor lighting in the main office.
13. Please inform each factory and tell them to cancel all other appointment.
14. Steven would rather attend the earlier conference—he is busy the week after.
15. Please note that a purchase order must be completed for all purchases over £ 65.

Section B

Listen to the audio clip 2-2 and type the following longer and more difficult sentences.

clip 2-2

1. It's important that you take the time to craft a good memo so your message comes across how you want it to.
2. When communicating with colleagues and other internal stakeholders through a memorandum, it is important to use the proper format.
3. The purpose of a memo is usually found in the opening paragraphs and is presented in three parts: the context and problem, the specific assignment or task, and the purpose of the memo.
4. I think it would be a good idea to include some of our overseas clients on the guest list for the formal event we are planning as part of the company's 30th anniversary celebrations.
5. Management is also aware that this might impact those sales reps and this news may impact them adversely.
6. We are prepared to relieve quota on those sales reps who can demonstrate those customers they have received commitments from their managers.
7. On 15 November, Ms. Anita Trosborg, Design Director of the ICD Design Company of Copenhagen is paying a visit to our offices.
8. Mary has agreed to give a talk on international recruitment, and she'll take three colleagues along with her.
9. She is giving a PowerPoint presentation and just needs a screen—she'll take her laptop with her.
10. There will be two one-day conferences this year—one in Edinburgh on September 2 and the other in London on September 12.
11. All training materials will be provided but students will be expected to do homework and preparation outside working hours.

12. They will be given an informal oral test during the first week in May, so that we can decide which of the classes is best for them.
13. Following a meeting yesterday afternoon with the Workers' Council, we have come to an agreement about employee overtime.
14. Complete purchase orders should be passed to Steve William for agreement of terms of payment with the supplier, and then sent to the Ottawa Office for final approval.
15. The Board is thinking of installing an automatic clocking-in system in the offices of each division.
16. Although we have reluctantly decided to raise your costs, be assured that we have done everything in our power to keep that raise to a minimum.
17. There have been a number of comments about the amount of time being wasted with extended lunch breaks in our company.
18. As a result of the productivity survey carried out in the factory, more rapid and efficient ways of operating are now being applied.
19. We therefore propose to pay a month's extra salary to any person who, in the management's opinion, has put forward the most practical suggestion to improve a particular office routine.
20. It has become increasingly important to use the Internet as a tool to communicate with our target audience.

Section C

clip 2-3

Listen to the audio clip 2-3 and type the following paragraphs.

1. I am writing to inform you of a baby shower we're planning for Kelsey Johnson, before she leaves on maternity leave. The party will be held on the 2nd floor, in conference room 2B, on Friday, October 30, 2021. Pizza will be provided by the company. RSVP to John Smith by Wednesday, October 28, and please include in your RSVP any dietary restrictions.
2. Please be advised that the Human Resources Department will provide First Aid training for all interested personnel. The dates will be June 20, 2021 and June 30, 2021. We are excited about the benefits of offering this important training and we are confident it will be of great value for our company. You may confirm your interest by calling our Human Resources Department no later than June 15, 2021. Thank you in advance for your cooperation.
3. Market research and analysis show that the proposed advertising media for the new fall lines need to be changed. Findings from focus groups and surveys have made it

apparent that we need to update our advertising efforts to align them with the styles and trends of young adults today.

Section D

Listen to the audio clip 2-4 and type the following memo sentence by sentence.

clip 2-4

A

To: All Employees in the Procurement Department
From: Mr. James French, Assistant Manager, Staff Training and Development
Date: August 22, 2021
Subject: Mandatory Training for New Software

You are all aware of the company's recent adoption of a new supply chain management software. The company invested in the new application to improve communication with our vendors, enhance order tracking and reduce delays to the barest minimum to save cost and boost efficiency.

We will hold a training workshop to familiarize department staff with the new software on August 25, 2021. We hope this training will allow everyone to make a smooth transition to the new application. All departmental staff must attend the training event.

Listen to the audio clip 2-5 and type the following memo sentence by sentence.

clip 2-5

B

To: Fintech Sales Team
From: Janet Underwood, Head of Sales
Date: May 20, 2021
Subject: Sales Quota Achievement

I am writing to congratulate you on the commendable efforts and energy you put into delivering on your team's sales quota for the last quarter.

Your team showed exemplary product knowledge, customer service, negotiation skills and collaboration that is worth emulating by other teams and departments in the company.

Thanks for your dedication and commitment to excellence. We will send your bonus checks and letters of commendation by the end of the week.

Congratulations on this achievement!

clip 2-6

Listen to the audio clip 2-6 and type the following memo sentence by sentence.

 **C**

To: All Staff
From: Leila Smith, General Manager
Date: February 18, 2021
Subject: Recurring Data Security Issues

It has come to my attention that the company has experienced a series of data breaches in the last two weeks because of the recent change in our firewall security system. I know some of you have lost files on your workstations and there has been at least an accidental leak of sensitive company information. We are taking steps to address the issue to prevent further data security lapses.

In the meantime, we are switching to a temporary data management system until our engineers and external consultants can fix the problem. We welcome comments and suggestions on how to solve this problem so we can get back to delivering results for our customers.

Thanks for your understanding.

clip 2-7

Listen to the audio clip 2-7 and type the following memo sentence by sentence.

➢ **D**

To: All Employees
From: James Paul, VP, Production Department
Subject: Periodic Factory Maintenance
Date: September 21, 2021

I' m writing to inform you that the next periodic factory maintenance will start on September 27, 2021, and last for the next three weeks.

As we continue to create new products and increase our production volume, we feel it is necessary to keep our facilities in the best conditions to ensure conducive working conditions for our staff, meet customer expectations and hit our revenue targets.

During the periodic maintenance, we will shut down one factory each week and increase the shifts at the two operational facilities to meet our production quotas. We have made adequate arrangements for overtime bonuses and already discussed with heads of departments and team leaders on ways to maintain staff efficiency and productivity within the period of maintenance.

If you have concerns or questions regarding the scheduled factory maintenance, kindly contact the Human Resources Department. Meanwhile, we are sorry for any inconvenience that may result from this operation.

Thank you for your cooperation.

Notes

accounting	*n.*	会计
implement	*v.*	实施
coverage	*n.*	保险项目
Board	*n.*	董事会
forthcoming	*a.*	即将到来的
stakeholder	*n.*	利益相关方;参与者
anniversary	*n*	周年纪念日
quota	*n.*	定额;指标
relieve	*v.*	解除;减轻
sales reps		销售代表
adversely	*ad.*	不利地
demonstrate	*v.*	展示
commitment	*n.*	承诺;奉献
Copenhagen		哥本哈根
recruitment	*n.*	招聘
Edinburgh		爱丁堡
council	*n.*	委员会
overtime	*n.*	加班
supplier	*n.*	供应商
Ottawa		渥太华
automatic	*a.*	自动的
division	*n.*	部门
reluctantly	*ad.*	勉强地
routine	*a./n.*	日常的;常规
baby shower		迎婴派对(在孩子出生前举办的特殊派对)
maternity	*a.*	产妇的
RSVP		请赐复(用于请柬结尾请求答复)
dietary	*a.*	有关饮食的
restriction	*n.*	限制
first aid		急救
align	*v.*	排整齐;使一致
procurement	*n.*	采购

mandatory	*a.*	强制性的
adoption	*n.*	采用
supply chain		供应链
vendor	*n.*	小贩;卖方;供应商
familiarize	*v.*	(使)熟悉;(使)了解
transition	*n.*	过渡
commendable	*a.*	值得赞扬的
quarter	*n.*	季度
exemplary	*a.*	典范的
collaboration	*n.*	合作
emulate	*v.*	效仿;同……竞争
bonus	*n.*	奖金
commendation	*n.*	赞扬;奖励
recur	*v.*	反复出现
breach	*n.*	(对法规等的)违背;违犯
firewall	*n.*	防火墙
workstation	*n.*	(计算机)工作站
leak	*v./n.*	透露(秘密信息)
lapse	*n.*	过失;疏忽
switch	*v.*	转向
temporary	*a.*	临时的
consultant	*n.*	顾问
VP (=vice president)		副总裁
periodic	*a.*	周期的
maintenance	*n.*	维护
conducive	*a.*	有益的;有助于……的
revenue	*n.*	收益
operational	*a.*	操作的;运转的

2.2 Telephone Messages 电话留言

Part One: Warm-up Activities

In Part One, you will practice typing sentences, paragraphs and passages dealing with telephone messages. Firstly you are supposed to read aloud and to identify the new words listed in Section A, and then try to practice typing the sentences, paragraphs and passages in Section

B，C，D respectively under the teacher's guidance. While you are typing，please mark out the time you spend on each section and compare your results with your classmates.

Section A Vocabulary work

transfer	*v.*	转接
available	*a.*	有空的
put...through		接通……
afterwards	*ad.*	以后

Section B Sentence practice

1. Good morning. Sales Department. What can I do for you?
2. Could I speak to Mr. Liu?
3. When do you expect him back?
4. Hold on for a moment，please. I'll transfer your call.
5. I'll tell her as soon as she is back.
6. Please tell him to call the doctor's office at 8642368.
7. I'm sorry，but he is not in at the moment.
8. Can you tell me when he will be in?
9. Can I take a message for him?
10. May I have your name please?

Section C Paragraph practice

1. Paul，Jack called to say he would come to the company to discuss the sales contract the next day. If inconvenient，you are expected to call him back. His number is 0435657689.
2. Lin Feng called to say he would go to Pudong Airport to pick you up at 2 p.m. next Monday. You are expected to wait at the exit.
3. Mr. White called from Paris. It is urgent：there's been a problem about the design of product number 132C—that's the one for the French market. Please call him back at 0033589847 till tomorrow evening.

Section D Passage practice

➤ Passage 1

（A：Assistant B：John Smith）

A：Good afternoon，Sales Department. May I help you?

B：Could I speak to Mr. Bush，please?

A: I'll see if he is available. Who shall I say is calling, please?

B: John Smith.

A: Hold the line, please. Mr. Bush is in a meeting with the Managing Director at the moment, I'm afraid. Can I help you?

B: Well, I want to discuss with him the new contract we signed last week.

A: I don't think the meeting will go on much longer. Shall I ask him to call you when he is back?

B: Yes, that would be fine.

A: Could I have your name again, please?

B: Yes. It's John Smith.

A: OK. You'll be hearing from Mr. Bush later in the morning then, Mr. Smith.

B: Thank you for your help. Good-bye.

A: You are welcome. Good-bye.

➢ Passage 2

(A:Operator B:Mr. Ma)

A: Operator.

B: Hello. I'd like to make a call to Australia.

A: You can call directly if you like.

B: Oh, can I?

A: Yes, please.

B: Could you please tell me the international code and the country code for Australia?

A: Yes. The code for Australia is 61 and then dial the city code and the number. Start with the international code which is 00.

B: What time do the special rates apply?

A: Between six in the evening and eight in the morning, sir.

B: I wonder if I can charge this call to my hotel room.

A: Certainly. Tell me the room number and your name, please?

B: This is Mr. Ma in room 215.

A: OK, Mr. Ma. I'll tell the front desk clerk the charge afterwards.

B: Thanks very much for your help.

A: You're welcome. Bye.

B: Bye.

➢ Passage 3

(A:Operator B:Mr. Shaws C:Secretary)

A: Hello. International Sales.

B: Hello. This is Mr. Shaws.

A: Yes, Mr. Shaws. Who would you like to speak to?

B: I'd like to speak to Mr. Mathews.

A: Fine. Hold on for a moment, please. I'm connecting you now.

C: Hello. Mr. Mathews' office.

B: This is Mr. Shaws, calling from England. Could I speak to Mr. Mathews?

C: I'm afraid Mr. Mathews isn't available. He's gone to Hong Kong on business for a few days.

B: When do you expect him back?

C: He'll be back on Friday afternoon. Is it urgent?

B: Yes.

C: Can I take a message for him?

B: Yes, please. Will you tell him that we've just received your sample of the new assembling coffee table, and we're quite happy with it.

C: Sure. It's very kind of you to say so. Can we expect an order from you?

B: That's why I'm making the call. Please tell Mr. Mathews we're quite happy with the quality and design of the table, but the price is too high. We need some negotiation on it.

C: OK, Mr. Shaws. Anything else?

B: One more thing. Please inform Mr. Mathews that I won't be able to get to your company that early this Saturday because of the rail strike. It will probably be afternoon before I arrive.

C: No problem. I'll give him the message.

B: Thanks.

C: You're welcome. Goodbye.

Part Two: Audio-typing

Now you are going to listen to the recording. Do not refer to your textbook while you are listening. Then you are supposed to listen to each section sentence by sentence once again and type what you hear at the same time.

Section A

Listen to the audio clip 2-8 and type the following short and simple sentences.

clip 2-8

1. Hello, 84715168.

2. Hello. Is Mr. Smith in, please?
3. May I speak to Mrs. White, please?
4. I'd like to speak to Mr. Lee, please.
5. Who do you wish to talk to?
6. May I know who's calling, please?
7. Who's that?
8. Hello, this is Brown speaking from ABC.
9. Hello, is that Mr. Wang's office?
10. You have the wrong number.
11. Is he in?
12. Hold the line, please.
13. I'll find out if he is available.
14. Are you still there?
15. Mr. Wilson, there is a call for you.
16. Sorry, he's not available right now.
17. Would you like to leave a message?
18. Could you possibly ask him to call me back?
19. Please tell him to phone 68197056.
20. Please tell her Carol called.

Section B

clip 2-9

Listen to the audio clip 2-9 and type the following longer and more difficult sentences.

1. My name is Tony Smith, Shanghai Hotel Room 2107. My phone number is 65678900.
2. Please make a remittance of 1,500 Yuan for the books you've ordered. The postage is included.
3. Operator, we were cut off. Could you reconnect me, please?
4. Mr. John Green, our general manager, would like to call Mr. Zhang on June 3 at 2 p. m. about the opening of a sample room there.
5. Sorry. Something is wrong with the phone. It's not clear. Please repeat that.
6. This is the Singapore operator. Would you connect me with Mr. Lee in the International Department?
7. My wife and I can't come to the phone right now, but if you'll leave your name and number, we'll get back to you as soon as we're finished.
8. Hi, I'm not home right now but my answering machine is, so you can talk to it instead. Wait for the beep.

9. Hi, this is George. I'm sorry I can't answer the phone right now. Leave a message, and then wait by your phone until I call you back.
10. Hello. Overseas operator. I'd like to make a collect call to Japan.
11. Country code 81, area code 138, and the number is 8648972.
12. The charges vary according to the types of call you make. The cheapest is a station-to-station call, then a person-to-person call.
13. I'm sorry, but he has a visitor right now. Could you hold a little longer? Or shall I put you through to his secretary?
14. Hi, this is John. I'm not available to take your call, but please leave your name, number and a brief message. I'll get back to you as soon as possible.
15. Hi, Mr. Zhang. This is Roland. Miss Lee telephoned. She asked me to ask you if you would be able to meet her today at 3:30 p.m.
16. I'll just find out if he is available, sir. Hello, are you still there? Unfortunately, Mr. Chen is not available at the moment. Would you like me to put you through to Mr. Li?
17. Hello, Mike? It's Susan speaking. The line was very busy, and it took me quite a long time getting to you.
18. Operator. The connection was bad. There was an echo and I couldn't hear well. Will you put me through again?
19. I just came back about a week ago, I tried to contact you by phone several times, but you were not in.
20. Hello. Yes. This is Mr. Kent of Oriental Ltd. We are interested in the carpets advertised by you and I think the carpet will find a ready market in our country.

Section C

Listen to the audio clip 2-10 and type the following telephone messages sentence by sentence.

clip 2-10

1.

TELEPHONE MESSAGE

To: Louise Paulson

From: Paul Jackson

Telephone Number: 9793268965

Message: Ring back to him about the order they placed last Friday. They have to make some changes to the order. It is urgent.

Taken by: Roy

2.

FLIGHT RESERVATION FORM

NAME:Danny Randall　　DATE:May 20

FROM:Shanghai　　DEPARTURE TIME:9:00 a.m.

TO:Chicago　　ARRIVAL TIME:12:15 p.m.

FLIGHT NO. Z254

CONFIRMATION NO. TA145

3.

MESSAGE

From:Mike Park

To:Carla Davis

Date: 7th April　　Time: 10:10 a.m.

Message:Meeting place with Adriana changed from Grappa's to Cafe continental. Be there at 9:00 p.m.

Action: Call back if you have any problems 016997207743.

Signed: Frank Wu

Section D

clip 2-11

Listen to the audio clip 2-11 and type the following telephone conversation sentence by sentence.

➢A

(A:Receiver　B:Miss Zhang)

A: United Development Corp. May I help you?

B: I'd like to speak to Mr. Smith, please.

A: Who shall I say is calling, please?

B: This is Miss Zhang from ABC Corp.

A: I'm sorry, Miss Zhang, but Mr. Smith is not in at the moment.

B: When will he come in, do you know?

A: I suppose he won't be in until 11:00.

B: May I leave a message?

A: Certainly.

B: Please ask him to give me a call as soon as he returns. He has my number.

A: Very well, Miss Zhang, I'll do that.

B: Thank you. Goodbye.

Listen to the audio clip 2-12 and type the following telephone conversation sentence by sentence.

clip 2-12

➢ B

(A: Mr. Smith B: Melva Miller)

A: (On the phone) Hello? Smith here.

B: Oh, Mr. Smith, my name is Melva Miller. You don't know me, but I'm a friend of Mike Black.

A: Oh, yes?

B: When I told Mike I was coming to live here he gave me your name, and suggested that I give you a ring. I was wondering if you could give me some advice.

A: I'll be pleased to if I can. What can I do for you?

B: Well, I'm looking for a place to live in. Mike thought that as you're an estate agent you might know of something suitable.

A: Yes, I think I can help you. Why don't you come round and see me? Do you know where my office is?

B: Yes. I've got the address.

A: Good. Where are you now?

B: I'm at the post office.

A: Oh, well, that's just a few minutes' walk from my office. Come round and see me now.

B: Thank you very much, Mr. Smith.

A: Not at all.

Listen to the audio clip 2-13 and type the following telephone conversation sentence by sentence.

clip 2-13

➢ C

(A: Assistant B: Bill Patten)

A: Hello, thank you for calling ABC. This is Tracy speaking. May I help you?

B: Hello. I would like to speak to your director of Human Resources, Mrs. White, please.

A: Just a moment. I'll check to see if she is at her desk. May I tell her who is calling?

B: This is Bill Patten from Mcford Insurance. I'm calling in regards to our meeting next

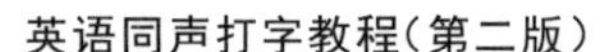

Tuesday.

A: Thank you, Mr. Patten. Can you please hold for a moment? I'll check to see if she is available.

B: No problem.

A: I'm sorry, Mrs. White is away from her desk. She has already left for lunch. Would you like to leave a message for her?

B: Yes, please have her return my call when she returns to the office. It's best if she can get in touch with me before 3 p.m. today. She can reach me at my office number, 6358766.

A: I'm sorry, I didn't quite catch that. Could you please repeat the number?

B: No problem. My office number is 6358766. Tell her to ask for extension 16.

A: I'm sorry, Mr. Patten, just to confirm. Your name is spelled P-A-T-T-E-N. Is that correct?

B: Yes, and I represent Mcford Insurance.

A: I will make sure Mrs. White receives your message and returns your call before 3 p.m. this afternoon.

B: Thank you very much.

clip 2-14

Listen to the audio clip 2-14 and type the following telephone conversation sentence by sentence.

➢ D

(A: Sandy B: Sam Darcy)

A: Hello, Pasaden Inn. This is Sandy. How may I direct your call?

B: I'd like to speak to someone about reservations.

A: I can help you with that. What date would you like to make a reservation for?

B: We'll be arriving on May 12th, but I would like to make reservations for a penthouse.

A: Oh, I'm sorry, sir. I only handle bookings for our standard rooms. The person you need to speak with is Tony Parker. He makes all the arrangements for our executive accounts. Unfortunately, he's not here right now. Can I take your name and number and have him get back to you?

B: When do you expect him back in?

A: He'll be out all afternoon. He might not be able to return your call until tomorrow. Will that be alright?

B: Yes, I suppose. My name is Sam Darcy. He can contact me at 6608433236.

A: Could you please spell your last name for me?

B: Sure. It's D-A-R-C-Y.

A：Okay. Mr. Darcy，and your phone number is 6608433233?

B：That's 3236.

A：I' II have Tony call you first thing tomorrow morning.

B：Thank you very much.

Notes

collect call		对方付费电话
station-to-station call		叫号电话
person-to-person call		叫人电话；传呼电话
Human Resources		人力资源（部）
estate agent		房地产经纪人
reservation	*n.*	预定
penthouse	*n.*	顶层豪华套间
standard room		标准间
executive account		高级客户
extension	*n.*	电话分机

2.3 Minutes 会议纪要

Part One：Warm-up Activities

In Part One，you will practice typing sentences，paragraphs and passages dealing with minutes. Firstly you are supposed to read aloud and to identify the new words listed in Section A，and then try to practice typing the sentences，paragraphs and passages in Section B，C，D respectively under the teacher's guidance. While you are typing，please mark out the time you spend on each section and compare your results with your classmates.

Section A Vocabulary work

malfunction	*n.*	（机器等的）故障；损坏
shortage of funds		资金短缺
maliciously	*ad.*	恶意地
a new product launch		新品发布
live broadcast		直播
Administrative Department		行政部门
cash flow		现金流

adjourn v. 休会

Section B Sentence practice

1. This is the model of our new product which is being designed.
2. Ms. Shang, you should contact the client who left his information at the exhibition yesterday.
3. If this malfunction of these machines cannot be repaired in time, we will have to delay the launch of the product.
4. In view of this emergency, we should hold an emergency press conference.
5. The factory replied that they expected to complete our order next month.
6. This is the new product that we have developed this time. Please give your opinion according to the market demand and competitive products.
7. The project manager is saying that the project is being held up because of the shortage of funds.
8. The supplier replied that the goods would be shipped within five days.
9. It is said that competitors are raising prices substantially.
10. Manager Liu is in full charge of the customer's visit to the factory this time.

Section C Paragraph practice

1. Wuhan Electronics Co., Ltd. will hold a new product launch at 7:00 p.m. on May 8, 2021, and will show the newly developed products and their functions to the public through a live broadcast.
2. In the past year, the products have been well developed and sold. This is inseparable from the efforts of all staff, and also proves that in today's era, market demand promotes the development of technology, and products should keep pace with the times to stimulate consumption.
3. According to the feedback from the Administrative Department, in order to provide a good working environment for everyone, the company will add five electric massage chairs on each floor for employees.

Section D Passage practice

Sunday Top Meeting

June 10, 2021

Electronic Co., LTD held a senior management meeting at 10:00 a.m. on June 10. All the

department managers attended the meeting.

The meeting proposed that according to the competitor's product released the day before yesterday, there are many similarities with upcoming online products, so the product information may have been leaked. All department managers must conduct investigation, and the Public Relations Department needs to carry out the public relations for the new product release, while the R&D Department needs to change the products' design and save the relevant data. The Sales Department needs to recommend other products to customers and inform them of the delay of the new product optimization.

The meeting adjourned at 11:30 a.m.

Annual Meeting

Time: December 30, 2020

Place: Greenlet Hotel

Chairman: Mr. Jerry, Chairman of the Wuhan Electronics Co., Ltd.

Minutes keeper: Shery

Wuhan Electronics Co., Ltd. held its annual meeting in Greenlet Hotel at 4: 00 p.m. this afternoon. First of all, the chairman made a speech at the opening ceremony. He summarized the annual work of the company, looked forward to the next year's work objectives, and gave his best wishes to the employees.

This was followed by a speech by the general manager on the total annual sales, and a brief introduction by the Technical Department on the product development that is expected to be carried out in the coming year. The department managers gave a brief introduction to the annual work of their respective departments.

The meeting adjourned at 7:00 p.m.

Part Two: Audio-typing

Now you are going to listen to the recording. Do not refer to your textbook while you are listening. Then you are supposed to listen to each section sentence by sentence once again, and type what you hear at the same time.

Listen to the audio clip 2-15 and type the following short and simple sentences.

1. The meeting will be held at seven o'clock.

clip 2-15

2. The board of directors disapproved of the plan.
3. The company has a very good reputation in the industry.
4. This project has been praised by the board for its unique creativity and rigorous program.
5. Considering the limited cash flow at present, the chairman chose to invest conservatively.
6. At the meeting, the general manager pointed out the difficulties the company was facing at the moment, but there is not a good way to solve them.
7. The chairman made it clear at the meeting that professionals should be hired, no matter how much money is spent.
8. It was decided at the meeting that the branch must develop its business in line with the local economy.
9. The manager of the Human Resources Department gave his opinion on the problem of large staff turnover in the company at the meeting.
10. Since the factory we are cooperating with has not delivered the goods on time for a long time, Mr. Smith asked whether we need to choose another in front of all the directors today.

Section B

clip 2-16

Listen to the audio clip 2-16 and type the following longer and more difficult sentences.

1. The Chairman has requested that a board meeting be held on Tuesday, April 5, 2021 at 2:00 p.m.
2. Business Celebration Banquet will be held at Hall 5 of Green Hotel, 323 Evergreen Road, on Friday, April 20, 2021, at 7:00 p.m.
3. Linda is showing our company's sales for this quarter, which are down significantly from the previous two quarters.
4. According to the investigation, the function of this detector developed by us is relatively rare in the market, so its market competitiveness is absolutely strong.
5. The personnel manager pointed out that the main problem affecting the turnover of the company is the imperfect welfare package.
6. At the meeting, the sales manager refused to place all the blame for the decline in sales on his department.
7. Today's meeting is mainly to solve the unexpected situation of the goods in the process of long distance transport.
8. After reviewing the sales and profits for the quarter, the chairman talks to the sales

and production managers individually at the end of the meeting.

9. The issue was made more difficult by a temporary price change by the acquiring company, and the atmosphere at the meeting was frozen.
10. After a discussion by the department managers, it was finally decided that Miss White would lead Liu Fang and Wang Hua to attend the International Electronic Exhibition in Hungary.
11. After reading the detailed plan, the directors discussed and finally agreed to form an inspection team.
12. The sales manager pointed out that next month's sales need to be increased by 10%.
13. The factory reported that the shipment could not be delivered on time due to problems in the supply of raw materials.
14. Whether to relocate the head office to Beijing remains to be considered.
15. According to the financial statements of each branch in this quarter, the Xiamen branch has the most sales.
16. In this annual conference, the R&D Department demonstrated the new robots currently being developed and received unanimous praise.
17. Based on the analysis of the parameters of chips, screens, etc. and the current market demand, our company decided to send the Production Department to investigate market condition.
18. After the new product is released, we need change the pricing strategy.
19. Mr. Smith said that his company is accustomed to using CIF quotations and hoped that our company would give them preferential prices.
20. Miss White won the best interest for our company in the negotiation meeting.

Section C

Listen to the audio clip 2-17 and type the following paragraphs sentence by sentence.

clip 2-17

1. At the negotiation meeting held on May 5, 2021, the negotiators of the other side were aggressive, hoping to obtain the most favorable prices and concessions. But if we follow the price they requested, then our profit will be greatly reduced. This is really a tricky problem.
2. Due to the weather, the output of raw materials has decreased and the prices have risen. A large number of raw materials cannot be received according to the price of our order, so the current supply of raw materials is insufficient and cannot be delivered on time.
3. At this staff meeting, the administrative manager reported that the computers currently

used by the company are not sufficient for daily work. However, if the entire company replaces computers that meet long-term needs, it is estimated to cost 1.5 million dollars, which exceeds the financial budget of 1.3 million dollars.

Section D

clip 2-18

Listen to the audio clip 2-18 and type the following minutes sentence by sentence.

 **A**

MINUTES OF INVESTING ABROAD MEETING

Date: June 5, 2021, 7:00 p.m.
Venue: Greenlight Hotel Hall 5
Present: All Department Managers

First, the general manager made an analysis of the company's development prospects and investment plans as the opening remarks, and then the chief financial officer explained the company's current financial status and disposable cash flow.

Mr. Jack, the marketing manager, analyzed several investment plans proposed by the general manager on the current market development, and believed that it would be unwise to develop the electronics industry in small and medium-sized countries when the company's cash flow is not very ample. However, they can consider investing in the electronics industry in China, because economy of China is developing rapidly and the population base is large, and China has played an increasingly important role in the world in recent years.

Ms. Mary, the sales manager, said that China's development in Asia is indeed very rapid, but the company's disposable cash flow is currently limited. Is it prioritized to invest in countries near the head office?

Due to disagreements, the board of directors decided that Mr. Jack and Ms. Mary will make a detailed plan after the meeting, with detailed data and theories, which will be elaborated again at the meeting next week.

The meeting ended at 9:40 in the evening.

clip 2-19

Listen to the audio clip 2-19 and type the following minutes sentence by sentence.

B

Minutes of the Shareholder Meeting of Wuhan Electronics Co., Ltd.

Date: May 10, 2021, 3:00 p.m.

Venue: Meeting Room on the Third Floor of the Garden Hotel

Present: All Department Managers (Bu Shulin, Deng Wei, Hao Caijia, Mao Kai, Fan Changlin, Li Si, Liu Zhuanhu, Wang Bo)

On May 10, 2021, Wuhan Electronics Co., Ltd. held a shareholder meeting in the meeting room on the third floor of the Garden Hotel. Chairman Deng Wei presided over the meeting. At the meeting, two important things were considered: 1. Whether Mao Kai was responsible for the change of the chief executive officer of Wuhan Electronics Co., Ltd. 2. As the former chief executive officer was transferred to Hunan branch, Xu Nuona is now the chief executive officer of the company.

After deliberation by the board of directors, the above decision was unanimously passed.

Wuhan Electronics Co., Ltd.

Listen to the audio clip 2-20 and type the following minutes sentence by sentence.

clip 2-20

 **C**

Date: 10 June 2021

Venue: The Union Hall

Subject: Zhejiang Factory Regular Meeting

Present: All Workshop Directors

Discussion:

Due to the recent frequent rain, the cotton output has declined, and our current stock is only enough to meet the current order, and we cannot take any more orders. Do you have any good suggestions for this problem?

1. Report by purchasing manager

Manager Liu of the Purchasing Department distributed a material to everyone, which showed the rainfall in the Yangtze River Delta, the current status of the cotton fields, the recent changes in cotton production and prices, etc. But our factory mainly uses cotton fabrics as the main goods, which is a thorny issue for us.

As summer is about to enter, the demand for cotton fabrics will rise sharply. Since the output of cotton has fallen and prices have risen, we can consider other replacements. For example, silk fabrics, which are currently popular on the market, are more readily available than cotton, and their output is relatively high.

2. Other people's opinions

Mr. Fan thinks this decision is a bit risky, because the factory has been making cotton textiles for many years, and the machines and technicians are quite skilled in this work.

But Manager Li believes that although this method is risky, it can be tried. As for the machine, we can order it immediately. In terms of technology, we can arrange for some workers to learn first without affecting the progress of other orders. At the same time, we can use our surplus cotton combined with local castor to make cotton and linen textiles to improve our market competitiveness. At the same time, we must strengthen the training of employees.

3. Resolved

The two parties disagreed, so we finally decided to conduct market inspections of silk, cotton and linen. In addition, we need a return visit to all our customers, stating that due to the decrease in raw material output and the increase in prices, we will not accept new orders in a short time.

4. Data of next meeting

A meeting will be held here at 10 o'clock next Wednesday morning to discuss whether to use other material textiles instead of cotton fabrics. Manager Liu is asked to lead the staff to do the corresponding market research. Please join us on time next Wednesday!

The meeting adjourned at 9:00 p.m.

Notes

local economy		当地经济
staff turnover		员工流失率
significantly	*ad.*	显著地
detector	*n.*	探测器
welfare package		福利计划
acquiring company		收购公司
International Electronic Exhibition		国际电子展览会
Hungary		匈牙利
unanimous	*a.*	一致的
parameter	*n.*	参数
preferential price		优惠价
tricky/ thorny	*a.*	棘手的
disposable	*a.*	可自由使用的
shareholder meeting		股东大会
preside over		主持；负责
Yangtze River Delta		长江三角洲
cotton fabric		棉织品
castor	*n.*	蓖麻
linen	*n.*	亚麻

2.4 Reports 报告

Part One: Warm-up Activities

In Part One, you will practice typing sentences, paragraphs and passages dealing with business reports. Firstly you are supposed to read aloud and to identify the new words listed in Section A, and then try to practice typing the sentences, paragraphs and passages in Section B, C, D respectively under the teacher's guidance. While you are typing, please mark out the time you spend on each section and compare your results with your classmates.

Section A Vocabulary work

fulfill	*v.*	满足
sub-standard	*a.*	不合标准的
prevalent	*a.*	普遍存在的
permanently	*ad.*	永久性地
pandemic	*n.*	流行病
livelihood	*n.*	生计
catastrophic	*a.*	灾难性的
equivalent	*a.*	相等的
workforce	*n.*	劳动力
questionnaire	*n.*	问卷
distribute	*v.*	发布
COVID-19	*n.*	新型冠状病毒肺炎
collectively	*ad.*	总体来说
generate	*v.*	产生
bust	*a.*	破产的
motivation	*n.*	动机
perceive	*v.*	将……视为
eligible	a.	有资格的
precedent	*n.*	先例
reassessment	*n.*	重新考虑;重新评估
proportion	*n.*	份额;部分
attendance	*n.*	出席
hamper	*v.*	妨碍
envisage	*v.*	设想;展望
discrepancy	*n.*	不一致

WorkSet colour　　　　定工色牌

Section B Sentence practice

1. The term business report is very broad, and its scope extends up to almost all reports that are formally written to fulfill some business motive or objective.
2. The union suggested that sub-standard furniture and equipment should be replaced.
3. As requested by the Managing Director on 30 March, 2021, I have investigated the problems which have been raised concerning office health and safety.
4. Employers should offer various stress reduction programs to help employees manage stress because stress is prevalent in the workplace.
5. Newly appointed staff should be made aware of the company's safety regulation.
6. The purpose of this study was to determine the negative effects of stress on employees and the methods employers use to manage employees' stress.
7. It should be the responsibility of the Departmental Committee to instruct new staff on procedures for handling office equipment.
8. In the United States, 20% of firms with fewer than 500 employees closed permanently between March and August.
9. As requested, the purpose of this proposal is to describe and analyse the possible use of technology in the Marketing Department.
10. Our team will be able to work faster, more effectively and make greater use of technological innovations.

Section C Paragraph practice

1. The impact of the pandemic on livelihoods has been catastrophic, especially on those who have no savings, have lost their jobs or faced pay cuts. Working hours equivalent to 495 million jobs were lost in the second quarter of 2020—14% of the world's entire workforce.
2. This study was designed to determine the effects of stress on employees and to discover methods employers use to manage employees' stress. Sixty questionnaires were distributed to business employees in the Central Texas area, and the response rate was 78.3%. This section includes the Findings, Conclusions, and Recommendations.
3. Micro, small and medium-sized enterprises have been hardest hit by COVID-19. They are often collectively the largest employers in a country: in China, for example, they generate around 80% of employment.

Section D Passage practice

➢ Passage 1

Report on the Re-investment of This Year's Profits

Introduction

This report sets out to examine how the company should re-invest this year's profits. The areas under consideration are the purchase of new computers, the provision of language training courses and the payment of special bonuses.

Areas under consideration

• New computers

The majority of company computers are quite new and fast enough to handle the work done on them. Consequently, new computers would not be recommended.

• Language training courses

The company aims to increase exports, particularly in Spain and France. Therefore, language training courses would be an excellent idea for those employees who deal with overseas business partners and customers. In addition, training courses would increase motivation: staff would enjoy the lessons and perceive that the company is investing in them. Therefore, language training would be an option.

• Special bonus payments

Although special bonus payments would have a beneficial impact on motivation, they would have no direct effect on the company's operations. There are also potential problems concerning the selection of staff eligible for the payments and the setting of a precedent for future payments. Therefore, bonus payments would not be advisable.

Recommendations

It is felt that the best solution for both the company and staff would be to invest in language training. It is suggested that the company should organise courses in French and Spanish. Those employees who have contact with partners and customers should be assured of places but other interested members of staff should also be allowed to attend.

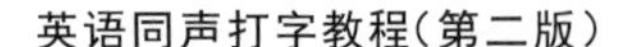

➢ **Passage 2**

Reassessment of Job Description

Introduction

This report sets out to use WorkSet colours to assess the accuracy of my job description as PR Officer and to suggest a number of changes.

Findings

It is clear that there is a difference between the way the company views the job and the reality as I perceive it. Firstly, a number of areas which demand a significant proportion of my time are not mentioned in the official job brief. The segment on the pie chart which provides the greatest cause for concern is the pink sector; this relates to my attendance at a number of meetings to which I can contribute little. Another significant area is the grey segment; this refers to the unscheduled time which I spend sorting out computer problems. I feel that these activities are hampering my core work. As can be seen from the pie charts, the time I spend actively working to meet the goals is less than envisaged.

Conclusions

The above discrepancies clearly indicate that my current job brief is inaccurate.

Recommendations

I would recommend that my official job brief should be updated using the WorkSet colours. It would also be valuable to consider the proportion of non-core colours in the pie chart and to investigate whether work in these areas could be carried out by someone else more suited to these tasks.

Part Two: Audio-typing

Now you are going to listen to the recording. Do not refer to your textbook while you are listening. Then you are supposed to listen to each section sentence by sentence once again and type what you hear at the same time.

Section A

Listen to the audio clip 2-21 and type the following short and simple sentences.

1. Most reports need at least an Introduction, Findings or Main Points and a Conclusion.

clip 2-21

2. Within each section, there should be clear paragraphs or bullet points.
3. Based on the findings and conclusions in this study, the following recommendations are made.
4. Sometimes if you need to do some in-depth research, the best way to present that information is with a research report.
5. Language training would help the company increase export sales to Spain and France.
6. Staff would enjoy the lessons and feel the company is investing in them.
7. Employees were generally satisfied with the current benefits package.
8. A study was made of all job-related illnesses during the past year.
9. Most employees complain about the lack of dental insurance in our benefits package.
10. Our benefits request system needs to be revised.
11. Improvements need to take place in Personnel Department response time.
12. Information technology improvements should be considered as employees become more technologically savvy.
13. Give priority to vacation request response time as employees need faster approval.
14. The purpose of this report was to analyse two portable computers.
15. There is no denying that these improvements will bring vast profits to the company.

Section B

Listen to the audio clip 2-22 and type the following longer and more difficult sentences.

clip 2-22

1. With continuous upgrading of public health strategies, many countries are now looking to ease measures such as travel bans.
2. The writer needs to ensure that the reader can identify the main points and information which is simply supporting the main points.

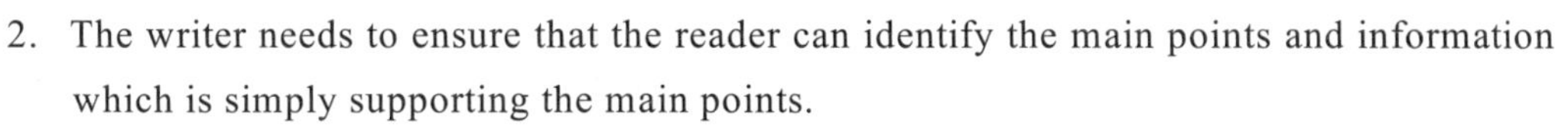

3. Annual report will typically round up a business year of progress and performance to let supervisors and team members know how the company did.
4. What is helpful to your team is a weekly report based on your progress in various projects and goals.
5. Sales-reports are also a great way to determine whether your strategies are working or if they need some tweaking in the future.
6. The Recommendations section could confirm which area should be invested in and outline ways the company might implement this proposal.
7. Even after offering a fair amount of salary to the staff on time, the annual turnover of the company has been between 55 to 62 percent every year.
8. While gathering data for this report, the human resource team has found that there is a

lack of support to the new mothers who want child care services so that they can come to work.

9. A lot of current employees mentioned how helpless they feel for not having child care system at home that can help them continue working without worrying much about the baby.
10. The interviewed staff expressed their concerns regarding confusing instructions and lack of clarity in work.
11. This report gives a summary of my meetings with various representatives of the Chinese tea industry and aims to give an impression of what it is like to do business in China.
12. In conclusion it would be advisable to buy the KD photocopier which is generally more advantageous in cost than its competitors.
13. Not only have we arranged promotional and advertising campaigns but we have also conducted market research via the Internet.
14. In our work we have used computer software and hardware which have already become obsolete and urgently need modernisation.
15. It seems obvious that the introduction of new technology into the Marketing Department will enable the company to gain huge profits.
16. The findings of this study indicated stress does negatively affect the work performance of employees.
17. In developed and emerging economies alike, the rapid shift to remote working is expected to yield long-term productivity gains.
18. Managers need to identify those suffering from negative stress and implement programs as a defense against stress.
19. COVID-19 has accelerated and broadened the Fourth Industrial Revolution with the rapid expansion of e-commerce, online education, digital health and remote work.
20. These shifts will continue to dramatically transform human interactions and livelihoods long after the pandemic is behind us.

Section C

clip 2-23

Listen to the audio clip 2-23 and type the following paragraphs sentence by sentence.

1. Therefore, stress cannot be considered just an individual issue because reduced job satisfaction and lower productivity has a direct effect on the company as a whole. From this study, it can be concluded that employers have realized the importance of managing stress in the workplace because of the wide variety of programs now offered

to manage stress.

2. Remembering the SARS epidemic, many countries in East Asia moved quickly, implementing a combination of travel bans, lockdowns and extensive testing with contact tracing, quickly targeting fresh outbreaks.
3. Today, many organizations and employees are experiencing the effects of stress on work performance. The effects of stress can be either positive or negative. What is perceived as positive stress by one person may be perceived as negative stress by another, since everyone perceives situations differently.

Section D

Listen to the audio clip 2-24 and type the following report sentence by sentence.

clip 2-24

 **A**

Resource Planning Manager: Assessment of Suitability for Home-Based Working

Introduction

The purpose of this report is to assess the suitability of my position as resource planning manager for home-based working.

Findings

My working pattern and that of my colleagues vary from week to week. During certain periods a large proportion of my time is spent doing fieldwork. This is followed by office-based work collating and recording the data collected. Once the results have been recorded, I proof-read the colour copies of all reports and maps.

As regards communication with colleagues, department meetings are held once a fortnight. At all other times, the individual members of the team communicate either face-to-face or by phone, depending on their location. Apart from official meetings, the same results can be achieved whether I am in the office or working elsewhere.

Conclusion

It is clear that I would be able to undertake the same duties while working from home for a large proportion of my time. Clearly, some days would need to be spent in the office for face-to-face communication with colleagues. It would also be necessary to use the technical facilities

of the office at times. However, in order to be able to work effectively from home, I would need to be provided with a networked computer and printer.

Recommendations

It would suggest that I should be given the necessary equipment to work partially from home for a trial period. After this time, further consultation should take place in order to reassess the situation.

clip 2-25

Listen to the audio clip 2-25 and type the following report sentence by sentence.

 **B**

A Sales Report

This report describes recent results in the Danish market, the reasons for them and suggests actions that we can take in the future to improve sales.

Overall, it has been a disappointing year, with sales falling by 30% compared with the previous year. However, this should be seen in the context of difficult trading conditions: everyone in the market is reporting decreased sales.

For us there are three factors causing our situation more difficult, they are:

—A new IKEA store was opened near Copenhagen four months ago, attracting business from other shops.

—We only have Italian brochures and customers would like them in Danish.

—Central, our biggest customer, has refused to order more lamps unless we increase their commission to 25%.

On a more positive note, the market seems to be recovering and consumers are spending again. We also have the prospect of a contract to supply lamps to the Chancery chain of hotels, which is about to refurnish twelve hotels here.

In order to take advantage of the improving market, I would like to make three recommendations:

1) that Danish brochures are made available as soon as possible.

2) that we increase the rate of commission to all our customers to 25%.

3) that we invest in some advertising in lifestyle and interior design magazines.

Please feel free to contact me about any of the above points.

Listen to the audio clip 2-26 and type the following report sentence by sentence.

clip 2-26

 **C**

Report on Staff Turnover in GHS Corporation

Submitted Aug. 8, 2021

Introduction

The human resources manager requested this report to examine the high turnover rate of employees at GHS Corporation. The information in this report was gathered by members of the Human Resources Department over three months. The five-member team analyzed administration records and working conditions, as well as interviewed staff. In this report, recommendations are made to minimize the high turnover rate among the staff working at GHS Corporation.

Background

GHS Corporation has been operating for 10 years. It employs 200 people, with most of the employees processing fees for insurance clients.

Findings

The most significant issue found by the HR team when interviewing staff was the lack of support to new mothers who require child care services to be able to come to work. Employees mentioned their frustration at not having an in-house child care system that could help them continue working.

Another issue mentioned by the staff was the lack of communication between employees and upper management. They expressed their concerns about receiving inconsistent and late instructions. They shared how they didn't know the main business objectives which led them to lose interest in the company and their jobs.

Conclusions

The main issues that we found were as follows:

1) lack of support to new mothers in regards to childcare.

2) lack of communication between the staff and upper management.

Recommendations

To address these two main issues, we recommend the following steps be taken:

1) an in-house childcare center can be established at minimal cost to GHS, encouraging

mothers to return to work.

2) each department should choose an employee ambassador to represent the interests of staff in management meetings. This ambassador can express concerns and relay outcomes to their teams to increase engagement.

clip 2-27

Listen to the audio clip 2-27 and type the following report sentence by sentence.

 **D**

Assessment on Possible Use of Technology in the Marketing Department

Purpose

As requested, the purpose of this proposal is to describe and analyse the possible use of technology in the Marketing Department.

Current use of technology

Up to now the members of our department have taken advantage of technological equipment, i.e., computers, in order to launch our products most successfully. Not only have we arranged promotional and advertising campaigns but we have also conducted market research via the Internet so as to meet our consumers' growing demands.

Technological improvements

In our work we have used computer software and hardware which have already become obsolete and urgently need modernisation. Therefore, the purchase of up-to-date programmes and equipment is of prime importance.

Benefits

There is no denying that these improvements will bring vast profits to the company. Our team will be able to work faster, more effectively and make greater use of technological innovations. Moreover, our company will be more likely to easily overcome fierce competition in the market.

Training

It seems obvious that our staff does not possess the knowledge of how to use new software. Thus, training on the use of modern programmes would be recommended as necessary.

Conclusion

To sum up, it seems obvious that the introduction of new technology into the Marketing Department will enable the company to gain huge profits. Our position in the market will be strengthened.

Listen to the audio clip 2-28 and type the following report sentence by sentence.

clip 2-28

 **E**

Report

Following your memorandum of 27 April, we carried out a small study of staff views in three selected departments to see how the arrangements of breaks had been working. I here summarize the results:

a) 65% office workers found the present break arrangements satisfactory.

b) 25% would be in favour of a shorter lunch break and finishing earlier.

It is too early to say definitely how many machines would be needed. But at least one for every divisional office seems a reasonable estimate.

I also asked my personnel officers about the saving of time. They think that an improvement in time-keeping could be made.

The staff's reaction to the idea was not very encouraging. In the survey we found out only 15% said they would be in favour of using clocking-in machines. If they had the choice, they would prefer not to use them.

You also asked for my views on how to deal with the union. I had a meeting with the chief union representative. I mentioned that in some departments the lunch break was lasting a lot longer than is actually allowed. The representative's answer was not very helpful. She said the union would always insist on the lunch break being left as it is. There is a point beyond which no negotiation would be possible without asking all the union members in the company their opinions.

In conclusion, it seems important to draw the Board's attention to possible difficulties which the rapid installation of clocking-in machines could bring. We need to discuss the problem a little longer and with more people before taking any action, it would seem.

Notes

dental	*a.*	牙齿的
savvy	*a.*	有见识的
priority	*n.*	优先权
deny	*v.*	否认
tweak	*v.*	轻微调整
emerging	*a.*	新兴的
remote	*a.*	远程的
yield	*v.*	产生收益
healthcare	*n.*	医疗保健
accelerate	*v.*	加速;加快
defense	*n.*	防御;抵御
ban	*n./v.*	禁止
fortnight	*n.*	两星期
IKEA		宜家家居
Danish	*n./a.*	丹麦语;丹麦语的;丹麦的
commission	*n.*	佣金
refurnish	*v.*	重新装修
clarity	*n.*	清晰;明确
obsolete	*a.*	过时的;淘汰的
ambassador	*n.*	大使
campaign	*n.*	运动;活动
clocking-in	*n.*	考勤

Chapter 3 Business Communication

商务沟通

3.1 Invitation and Establishing Business Relations 邀请和建立业务关系

Part One: Warm-up Activities

In Part One, you will practice typing sentences, paragraphs and passages dealing with invitation and establishing business relations. Firstly you are supposed to read aloud and to identify the new words listed in Section A, and then try to practice typing the sentences, paragraphs and passages in Section B, C, D respectively under the teacher's guidance. While you are typing, please mark out the time you spend on each section and compare your results with your classmates.

Section A Vocabulary work

leading exporter		主要出口商
textiles	*n.*	纺织业
handle	*v.*	经营
attached please find		随信附寄
catalogue	*n.*	目录
for your reference		供你方参考
owe your name and address to		从某处获悉贵方的名称和地址
Lagos Branch		拉各斯(尼日利亚旧都和最大港口城市)分公司
printed shirting		印花细布;印花衬布
under separate cover		另邮;另寄;在另函中

quotation	*n.*	报价
sample	*n.*	样品
tablecloth	*n.*	台布
workmanship	*n.*	工艺
in the market for		想要购买……
garment	*n.*	服装
avail oneself of		利用……
acquaint	*v.*	使……熟悉
business line		业务范围
enclose	*v.*	附寄
merchant	*n.*	商人

Section B Sentence practice

1. We have obtained your name and address from *China Daily*.
2. We are writing to you in the hope of establishing business relations with you.
3. We shall be pleased to enter into business relations with you.
4. We are a leading exporter in the business of textiles in North Africa.
5. Our company handles the import and export of agricultural products.
6. Attached please find our catalogue for your reference.
7. Please find enclosed our brochure illustrating all our products.
8. If you are interested in any of our products, please contact us.
9. We look forward to hearing from you soon.
10. We shall be pleased to cooperate with your firm.

Section C Paragraph practice

1. We owe your name and address to the Bank of China, Lagos Branch, through whom we have learnt you are exporters of Chinese textiles. Now we are interested in importing your printed shirting.
2. We are sending you under separate cover by airmail a copy of the latest catalogue. Please let us know if there are any items which are of interest to you and we will send you quotations and samples.
3. We sell Chinese tablecloths. They are of good quality and have fine workmanship. Chinese tablecloths are very popular in Europe. We would like to work with you to market them in Canada.

Section D Passage practice

➢ **Passage 1**

We owe your name and address to your branch in Nanjing, who has informed us that you are in the market for men's garments. We avail ourselves of this opportunity to write to you in the hope of establishing business relations with you.

We are handling both the import and export of garments. In order to acquaint you with our business lines, we enclose a copy of our Export List covering the goods you required at present.

It is our trade policy to trade with merchants of various countries on the basis of equality and mutual benefit to exchange needed goods. We hope to promote, through mutual efforts, both trade and friendship.

We look forward to receiving your first order.

➢ **Passage 2**

Thank you for your letter of September 6, 2020 showing your interest in doing business with us. We really appreciate your interest.

However, we very much regret that we are not in a position to establish business relations with you at present. We currently have another company as our agricultural products supplier in Suzhou. According to the contract conditions, we are banned from importing agricultural products from other companies.

As this is not an appropriate time to cooperate with your company, we would like to wait. But we will keep your letter on file and will get in touch with you when this contract expires.

We really hope that we have the chance to cooperate in the near future.

Part Two: Audio-typing

Now you are going to listen to the recording. Do not refer to your textbook while you are listening. Then you are supposed to listen to each section sentence by sentence once again and type what you hear at the same time.

Section A

Listen to the audio clip 3-1 and type the following short and simple sentences.

clip 3-1

1. Your early replies are highly appreciated.
2. We hope to hear from you soon.
3. We shall be very pleased to work with you on good terms.

4. We are one of the largest exporters in light industrial products.
5. We shall be glad to establish business relations with you.
6. We are interested in buying varied beans.
7. We await your early reply.
8. We highly appreciate your kind cooperation.
9. We are looking forward to your favorable and prompt reply.
10. Should any of the items be of interest to you, please let us know.
11. We enclose a catalogue and a price list for your reference.
12. We are looking forward to establishing business relations with you.
13. We hope you will handle this matter with the utmost care.
14. We look forward to your early reply with much interest.
15. Your wish of establishing relations with us coincides with ours.

Section B

clip 3-2

Listen to the audio clip 3-2 and type the following longer and more difficult sentences.

1. We owe your name and address to Italian Commercial Bank who has informed us that you are in the market for tablecloths.
2. On the recommendation of the Bank of China, we have learned the name of your firm.
3. Our products are well-known for their superior quality, unique materials and traditional craftsmanship.
4. Through the courtesy of the Chamber of Commerce in Tokyo, Japan, we have learned that you have been supplying the best quality food all over the world.
5. We learned from the Commercial Counselor of our Embassy in Ottawa that you deal in tablecloths.
6. We are looking forward to working with you on the basis of mutual benefit.
7. We are glad to send you this letter, hoping that it will be the prelude to our friendly business cooperation in the coming years.
8. We take the liberty of writing to you with a view to building up business relations with your firm.
9. Your letter of October 27, 2020 addressed to our Shanghai Branch Office has been passed on to us for attention and reply.
10. We got the information from our sales manager that you have the desire to cooperate with our firm in marketing our silk products.
11. We enclose a complete set of leaflets showing various products being handled by this corporation with detailed specifications and means of packing.

12. We have decided to appoint an agent to handle our export trade with your country.
13. Quotations and samples will be sent upon receipt of your specific enquiries.
14. If any of the items listed in the catalogue meets your interest, please let us have your specific enquiry, and our quotation will be forwarded without delay.
15. Your letter of Sept. 8 has been transferred to us for attention from our head office in Beijing.
16. We are China National Textiles Import and Export Corporation, with its headquarters in Beijing.
17. This is to introduce the Pacific Corporation as exporters of light industrial products having business relations with more than 70 countries in the world.
18. We are willing to enter into business relations with you on the basis of equality and mutual benefit.
19. As the item falls within the scope of our business, we shall be pleased to enter into direct business relations with you.
20. As the article lies within the scope of business of our branch in Nanjing, we have forwarded your letter to them for attention.

Section C

Listen to the audio clip 3-3 and type the following paragraphs sentence by sentence.

clip 3-3

1. We have your name and address from your branch in Tianjin, informing us that you are in the market for women's garments. We avail ourselves of this opportunity to write to you with a view to establishing business relations with you.
2. We are very well connected with all suppliers of agricultural products. Now, we are very interested in your beans and wheat, and feel sure we can sell large quantities of them if we get your offers at competitive prices.
3. We write to introduce ourselves as one of the largest exporters in China, of a wide range of machinery and equipment. We enclose a copy of our latest catalogue covering the details of all the items available at present, and hope some of these items will be of interest to you.
4. We are a privately-owned enterprise, handling the export of textiles. We wish to establish trade relationship with you by the commencement of some practical transactions. What's more, we enclose the catalogue and price list for your reference so as to acquaint you with our products available for export.

Section D

clip 3-4

Listen to the audio clip 3-4 and type the following letter sentence by sentence.

Dear Mr. James,

Dongfeng Imp. & Exp. Co., Ltd. has recommended you to us. We learn that you have been a leading distributor of sportswear in Canada for over 15 years.

We, Xiamen Tengda Industrial Co., Ltd., founded in 1998, specialize in the manufacturing and marketing of clothes and accessories for various sports, including biking, running and mountaineering, etc., and have now become a major supplier for many famous sports brands like Xtep and Li-Ning. Our products are well-known for its superior quality, novel design and fine workmanship. Attached you'll find a catalogue showing our current product line. You can also visit our website www.tengda.com for more information. Please let us know if you're interested. We shall be pleased to do business with you.

We look forward to your early replies.

Best Regards,

Laura

Sales Manager

Xiamen Tengda

clip 3-5

Listen to the audio clip 3-5 and type the following letter sentence by sentence.

Dear Sirs/Madams,

We have obtained your name and contact from China Council for the Promotion of International Trade (CCPIT). We have learnt that you have been dealing in children toys in Italian and European markets for many years, and are now interested in sourcing children toys from China. As a Chinese trading company in the same line, we're writing to you in the hope of establishing business relations with you.

We, Zhongtian Imp. & Exp. Co., Ltd is a state-owned trading company specializing in the import and export of children toys. Our products are extensively welcomed in the markets for its unique design and superior quality. After several years of development, we have established wide connections with not only suppliers of children products but also the dealers and retailers in China. We believe that our products will meet the needs of the consumers in your market.

For your convenience, we hereby attach our catalogue showing all our children toys currently

available. Please let us know the items you would be interested in. More information will be provided upon request.

We look forward to receiving your early reply.

Yours sincerely,

Sherry Chen

Sales Manager

Zhongtian Imp. & Exp. Co., Ltd.

Listen to the audio clip 3-6 and type the following letter sentence by sentence.

clip 3-6

 **C**

Dear Mr. Alan Blanco,

We learn from the magazine *Fashion* that you are a leading distributor of fashionable bags & suitcases in France and Europe. As bags & suitcases just fall within our business scope, we would like to take this opportunity to approach you and to see if there are any possibilities of business cooperation.

With nearly 20 years of experience in manufacturing and exporting bags & suitcases, we have grown to be a leading manufacturer in this field and have been supplying products for many famous brands, such as Rimowa and Lancel. Our products are very popular among young consumers thanks to their simple but fashion designs and traditional craftsmanship.

As we have a great variety of items available, we now attach our catalogue and price list for your convenience. You are also welcome to visit our official website to learn more about us. If you find any items that interest you, please let us know.

Thank you for your attention. We look forward to receiving your replies soon.

Best Regards,

Li Ping

Sales Manager

Guangdong Baige Bags & Suitcases

Listen to the audio clip 3-7 and type the following letter sentence by sentence.

clip 3-7

 **D**

Dear John,

This is Jenny from Zhejiang Jinhai Electric Appliance Co., Ltd. We met at the Canton Fair last week.

As you're looking for new suppliers of electric fans, we would like to introduce to you our company and our new product.

Zhejiang Jinhai is a large-size manufacturer of electric fans of various types. We have been engaged in the R&D, production and marketing of electric fans for more than 10 years. Recently, we have introduced to the market our new electric fan model JH40-3B which has been upgraded in many aspects. The new model has passed tests of the highest standards, and its superior quality is a guarantee for us to target the high-end market. Preliminary market feedbacks are positive and encouraging. For your convenience, we attach a brochure describing and illustrating this new model in details. We believe that the new model will sell well in your market. We shall be pleased to start doing business with you.

We look forward to receiving your comments or enquiries soon.

Regards,

Jenny

Zhejiang Jinhai Electric Appliance

Notes

light industrial product		轻工业产品
coincide with		与……相符
on the recommendation of		由……推荐
through the courtesy of		承蒙……介绍
Chamber of Commerce		商会
the Commercial Counselor		商务参赞
prelude	*n.*	开端;前奏
with a view to		目的在于……
be addressed to		寄给……;写给……
leaflet	*n.*	宣传册;传单
specification	*n.*	规格
upon receipt of		一收到就……
enquiry	*n.*	询价
forward	*v.*	发送;转寄;转交
head office		总部
fall within the scope of our business		……属于我方业务范围
offer	*n.*	报盘
privately-owned enterprise		民营企业;私企
distributor	*n.*	经销商
sportswear	*n.*	运动装;休闲服
specialize in		专营……

accessory	*n.*	配饰;配件
mountaineering	*n.*	爬山;登山
attached you'll find		随信附寄……
product line		产品线;产品系列
CCPIT		中国国际贸易促进委员会(China Council for the Promotion of International Trade)
source	*v.*	采购;从……获得
hereby	*ad.*	在此;凭此;据此
approach	*v.*	与……接洽;联系
manufacturer	*n.*	厂家;生产商
Rimowa		日默瓦(德国旅行箱品牌)
Lancel		兰姿(法国箱包品牌)
R&D		研究与开发
high-end market		高端市场
preliminary	*a.*	初始的
feedback	*n.*	反馈

3.2 Enquiries and Replies 询价与回复

Part One: Warm-up Activities

In Part One, you will practice typing sentences, paragraphs and passages dealing with enquiries and replies. Firstly you are supposed to read aloud and to identify the new words listed in Section A, and then try to practice typing the sentences, paragraphs and passages in Section B, C, D respectively under the teacher's guidance. While you are typing, please mark out the time you spend on each section and compare your results with your classmates.

Section A Vocabulary work

electric heater		电加热器;电暖气
illustrated	*a.*	有插图的
numerous	*a.*	许多的;很多的
nylon	*n.*	尼龙
at moderate prices		以适中的价格
artificial leather gloves		人造皮革手套
replenish	*v.*	重新充满;补货
revert to		回复;再次提及

couple with 加上;伴随
discriminating buyer 识货的买家

Section B Sentence practice

1. As we are in the market for men's leather boots, we should be pleased if you would send us your best quotation.
2. Would you please tell us the price of these electric heaters so as to help us make the decision?
3. We would appreciate your sending us an up-to-date price list for building materials.
4. We are interested in motorcycles in various sizes and please send us a copy of your illustrated catalogue with details of the prices and terms of payment.
5. Enquiries for carpets are getting more numerous.
6. We shall be glad if you will quote us the best discount for cash for this quantity.
7. We enclose our quotation sheet against your enquiry No. 28 and look forward to your confirmation.
8. We regret that the goods you enquire about are not available.
9. We look forward to your most favorable quotations.
10. Please let us know your lowest possible price for the goods we required.

Section C Paragraph practice

1. We learn from ABC Co. of New York that you are expecting nylon bed-sheets and pillow-cases. There is a steady supply here for the above-mentioned commodities of high quality at moderate prices.
2. Under separate cover, we are sending you a range of samples and when you have a chance to examine them, we feel confident that you will agree that the goods are both excellent in quality and very reasonable in price.
3. We have received your enquiry of 14th May and learned of your interest in our handmade artificial leather gloves. We are enclosing our illustrated catalogue and price list giving the details you ask for.

Section D Passage practice

➢ Passage 1

We were very pleased to receive your letter enquiring for our woolen sweaters. However, we regret to inform you that we are not in a position to cover your need for woolen sweaters. Once our supplies are replenished, we shall be pleased to revert to this matter. We are looking

forward to your more enquiries.

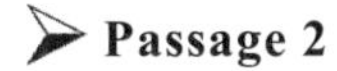

Passage 2

We are very pleased to receive your enquiry of June 20th and thank you for your interest in our products. A copy of our illustrated export catalogue will be sent to you today, together with a range of samples of the various leathers used in the manufacture of our gloves and shoes. We think the colors will be just what you want for the fashionable trade, and the beauty and elegance of our designs, coupled with the superb workmanship, should appeal to the discriminating buyers. Our representative, Mr. Black, will be in London next week and will be pleased to call on you with a full range of samples of our hand-made lines. He is authorized to discuss the terms of an order with you or to negotiate a contract.

Part Two: Audio-typing

Now you are going to listen to the recording. Do not refer to your textbook while you are listening. Then you are supposed to listen to each section sentence by sentence once again and type what you hear at the same time.

Section A

Listen to the audio clip 3-8 and type the following short and simple sentences.

clip 3-8

1. Please state your terms of payment and discount.
2. We should like to know if you allow discounts.
3. We are writing to enquire for your exhibited products.
4. Your early reply to this enquiry is requested.
5. We are looking forward to receiving your immediate reply.
6. Prices quoted should include insurance and freight to Liverpool.
7. Please let us know whether you are interested in such an order.
8. If your prices are competitive, we will place an order immediately.
9. We regret that the goods you enquired about are not available.
10. Your enquiry dated April 8th has been received with pleasure.
11. Attached please find our catalogue and price list for your reference.
12. Meanwhile, please inform us of payment terms, package and insurance.
13. We are thinking of getting a supply of beans.
14. Payment is to be made by irrevocable L/C at sight.
15. Our terms of payment are cash on delivery.

Section B

clip 3-9

Listen to the audio clip 3-9 and type the following longer and more difficult sentences.

1. We have seen your advertisement in *China Daily* and shall be grateful if you will send us details of iron nails.
2. Please send us your best offer indicating origin, packing, detailed specifications, quantity available and earliest time of shipment.
3. As we are in the market for men's leather gloves, we should be glad if you would send us your best offer.
4. We are desirous of your lowest quotations for sewing machine.
5. When quoting, please state terms of payment and time of delivery.
6. We shall be pleased if you will furnish us with your lowest quotation for the following goods.
7. If you supply goods of the type and quality required, we may place regular orders in large quantities.
8. We are on the look-out for the following items and should be appreciated if you would send samples of the same.
9. As the booklets you sent us were badly damaged in the mail, we would like you to mail us some more.
10. If the prices quoted are competitive and the quality is up to standard, we will place orders on a regular basis.
11. Please send us your latest catalogue with your best CIF London prices. We will also appreciate your telling us the approximate weight of each article.
12. We hope you will find our quotation satisfactory and look forward to receiving your order.
13. We welcome your enquiry of December 6, 2020 and thank you for your interest in our products.
14. The enclosed price list and illustrated catalogue will provide you with the details of the various types you are most interested in.
15. If your quotation is really reasonable and competitive, we will soon place a large order with you.
16. With regard to your enquiry for sewing machines, we wish to give the following in reply.
17. In reply to your enquiry of July 16, we are sending you our quotation together with various samples of leather boots closely resembling what you want.

18. We have much pleasure in enclosing a quotation sheet for our products and trust that their high quality will induce you to place a trial order.
19. We trust you will give this enquiry your immediate attention and let us have your reply at an early date.
20. We hope that your prices will be favorable and that business will materialize to our mutual advantage.

Section C

Listen to the audio clip 3-10 and type the following paragraphs sentence by sentence.

clip 3-10

1. We have just received an enquiry from one of our Japanese clients, who needs 10,000 metric tons of the captioned sugar and shall appreciate your quoting us your best price at the earlier date.
2. As there is a critical shortage of sugar in Japan, the goods should be ready for shipment as early as possible. Please be assured that if your price is acceptable, we will place an order with you right away.
3. We are in receipt of your samples with many thanks. We are satisfied with them. It would be highly appreciated if you could quote us your best price in USD per piece on CIF Montreal including 3% commission on your cotton blazers, Style No. BJ123.
4. For your information, the quality required should be superior white crystal sugar packed in new gunny bags of 200kgs each. Meanwhile, the goods should be surveyed by an independent surveyor as to their quality and weight before shipment.

Section D

Listen to the audio clip 3-11 and type the following letter sentence by sentence.

clip 3-11

A

Dear Thomas,

We have received your mail dated January 10th. Thank you for the message.

After carefully studying the catalogue attached in your mail, we find your marble countertop model M1305-B pretty interesting. If your price and terms would meet our expectations, you will have big orders from us.

Please quote us your best price, CIF New York, for a 20' FCL. We would also like to know your payment terms and earliest delivery date.

We look forward to your most favorable quotations.

Yours sincerely,

Abraham

Purchasing Manager

clip 3-12

Listen to the audio clip 3-12 and type the following letter sentence by sentence.

➢ B

Dear Sirs/Madams,

We know you from the Alibaba B2B platform, and are interested in buying your products.

Maxtor is a Chicago-based trading company specializing in the distribution of outdoor products, and has wide connections with many of the outdoor clubs and outdoor product retailers here. We are quite attracted by the outdoor backpacks that you show on Alibaba. However, before we decide to place an order, we would like to know more about the item especially in the following aspects:

1) What materials are used, and which colors are available for order?

2) How is it different from similar products in the market?

3) Is there any MOQ requirement per color, per model?

4) Can you give wholesale discounts, if we buy in large quantities?

To start with, we intend to buy 2,000 pcs as a trial. Please quote us your lowest price CIF Chicago. If your price is competitive, we will confirm the order immediately.

Thanks for your prompt reply.

Regards,

Jack Johnson

Brand Manager

clip 3-13

Listen to the audio clip 3-13 and type the following letter sentence by sentence.

➢ C

Dear Mark,

We have received your mail dated March 13th. We're pleased to know you're also dealing in dried fruits and have interest in doing business with us.

After we studied your catalogue, we find your dried blueberries particularly interesting. Please quote us your best price for a 20' FCL on the basis of FOB Vancouver. As there is an increasing demand for dried fruits recently, we would require an early delivery if your price is attractive to us. So please also let us know if you have the goods in stock, otherwise how much time you'll need for production.

Please give us your replies ASAP.

Regards,
Jerry Ma
Sourcing Manager

Listen to the audio clip 3-14 and type the following letter sentence by sentence.

clip 3-14

 **D**

Dear Liu,
We learn from your mail dated September 22nd that your company is engaged in the export of Chinese silk products, and is interested in establishing business relations with us.
As we are currently looking for suppliers of lady silk scarves, please let us know if you have such products in stock. If so, please let us have your sample book and quote us your lowest prices. Please also indicate your payment terms and lead time.
Meanwhile, as we are also interested in other silk products, please airmail to us your latest catalogue so that we can have a better picture of your product line.
We are looking forward to your favorable reply.
Yours,
James
Purchasing Manager

Notes

provided	*conj.*	假如;倘若
freight	*n.*	运费
place an order		订购……
irrevocable L/C		不可撤销的信用证
at sight		即期
cash on delivery		货到付现
iron nail		铁钉
origin	*n.*	原产地
on the lookout for		注意;留心
CIF		到岸价(Cost, Insurance and Freight)
article	*n.*	(单件)商品
prompt shipment		即期装运
materialize	*v.*	实现;使……具体化
mutual advantage		互惠互利
captioned	*a.*	标题下的;标题所说的

in receipt of		收到……
Montreal		蒙特利尔(加拿大东南部港口城市)
cotton blazer		全棉运动上衣
gunny bag		麻袋;麻布袋
FAS		船边交货价格(Free Alongside Ship)
marble countertop		大理石台面
a 20' FCL		20英尺集装箱包装的整箱货
Alibaba		阿里巴巴集团
B2B		企业对企业的电子商务模式;企业间电子商务(business to business)
backpack	*n.*	背包
item	*n.*	(单件)商品
MOQ		最小起定量(minimum order quantity)
wholesale	*n.*	批发
dried fruit		水果干
FOB		离岸价(Free on Board)
Vancouver		温哥华
in stock		有货;有库存
sourcing/ purchasing manager		采购经理
lead time		订货至交货的时间; 交付周期

3.3 Offers and Counter-offers 报盘与还盘

Part One: Warm-up Activities

In Part One, you will practice typing sentences, paragraphs and passages dealing with offers and counter-offers. Firstly you are supposed to read aloud and to identify the new words listed in Section A, and then try to practice typing the sentences, paragraphs and passages in Section B, C, D respectively under the teacher's guidance. While you are typing, please mark out the time you spend on each section and compare your results with your classmates.

Section A Vocabulary work

valid/firm	*a.*	有效的
be subject to		以……为准;以……为有效条件
counter-offer	*n.*	还盘
helmet	*n.*	头盔

in due course		适时地;在合适的时间
by separate post		另邮
regular purchase		长期购买
gross	*num.*	一罗（合144个）
softness and durability		柔软性和耐用性
justify	*v.*	证明……合理
come to terms		达成交易
review	*v.*	重新考虑

Section B Sentence practice

1. We are pleased to make our offer to you for the goods you enquire.
2. We take pleasure in making you an offer for the goods required.
3. The price we quoted is on FOB basis instead of CIF basis.
4. The offer is valid within three days.
5. The offer is firm, subject to our final confirmation.
6. The offer is subject to your reply reaching us before 8 March.
7. We believe you will find our offer competitive and satisfactory.
8. We look forward to receiving your order soon.
9. We hope to receive favorable replies from you soon.
10. Much to our regret, we are not able to accept your counter-offer.

Section C Paragraph practice

1. We have well received your enquiry dated March 22, in which you asked for our price for 1,500 pcs of helmets. Thanks for your trust in our products. As requested, we hereby quote you our best price for the goods required.
2. Thank you for your enquiry dated 17 July and, as requested, we are airmailing you, under separate cover, one catalogue and three sample books for our Australian royal wool blankets. We hope they will reach you in due course and will help you in making your selection.
3. We regret to say that we find your price rather high and we believe we will have a hard time convincing our clients at your price. Besides, there is keen competition from suppliers in Japan and China. You cannot ignore that. We find that we can obtain from another firm in China a price of 10% lower than that of yours.

Section D Passage practice

Passage 1

We are pleased to receive your enquiry of 10th January and enclose our illustrated catalogue and price list giving the details you ask for. We are also sending you by separate post some samples and feel confident that when you have examined them you will agree that goods are both excellent in quality and reasonable in price.

On regular purchases in quantities of not less than five gross of individual items we would allow you a discount of 2%. Payment is to be made by irrevocable L/C at sight.

Because of their softness and durability, our cotton bed-sheets and pillowcases are rapidly becoming popular, and after studying our prices you will not be surprised to learn that we are finding it difficult to meet the demand. But if you place your order not later than the end of this month, we would ensure prompt shipment.

Passage 2

We wish to thank you for your letter of Nov. 2, offering us 3,000 sets of 1.0P air-conditioners at USD 325 FOB Shenzhen.

We, however, regret to tell you that we find your quoted price is on the high side. Information indicates that some other suppliers are selling products of similar quality at a price 10%—15% lower than yours. We understand that the quality of your products is slightly better, but it does not justify such a big difference.

In this case, it is impossible for us to accept your offer, considering the market competition and our price strategy. Should you be prepared to reduce your price by 8%, we might come to terms. Please review your offer. We hope you will find our proposal most favorable.

Part Two: Audio-typing

Now you are going to listen to the recording. Do not refer to your textbook while you are listening. Then you are supposed to listen to each section sentence by sentence once again and type what you hear at the same time.

Section A

clip 3-15

Listen to the audio clip 3-15 and type the following short and simple sentences.

1. As requested, we're now making you the best offer for the goods required.

2. In order to reach an agreement, we need you to lower your price by 5% at least.
3. We hope you will find our offer most favorable.
4. We await your favorable replies.
5. We believe our prices are fixed at a reasonable level.
6. This quotation is subject to the fluctuations of the market.
7. We hope you would be satisfied with the above offer and await your order.
8. We look forward to your acceptance and first order.
9. Please note that this offer is firm, subject to your reply reaching us before Feb. 25.
10. We look forward to receiving your early replies.
11. The best we can do is to lower the price by 3%.
12. Business is possible if you can lower the price to USD 10.00 per piece.
13. If you would reduce your price by 5%, we might come to terms.
14. We would give you a discount of 5% if you increase your quantity to 10,000 sets.
15. We regret to tell you that the price you quoted is higher than the market price.

Section B

Listen to the audio clip 3-16 and type the following longer and more difficult sentences.

clip 3-16

1. Considering our good relationship and future cooperation, we will give you an extra 5% discount.
2. We note that you require payment by T/T before shipment, but we prefer irrevocable L/C for this first order.
3. The minimum quantity you require is too much for this market. If you can reduce the MOQ to 5,000 pcs, there is a possibility that we will place an order.
4. We regret to tell you that we cannot accept your offer as the price you quoted is above the average market price.
5. Please note that this offer is valid, subject to your reply reaching us before Oct. 15.
6. Thanks for your mail dated Oct. 5 in which you asked for our price for 10,000 pcs of promotional polo shirts.
7. We are pleased to hear that there is a great demand for our marble products in Australia.
8. Thank you for your mail dated May 8, inquiring about our marble floor tiles "Blue Bird".
9. Provided that we receive your order within the next ten days, we will give priority to it for prompt delivery.
10. Also, we have airmailed to you 3 pcs of samples for free. The tracking No. is

DHL8525635423. Hope the samples will arrive safely.

11. Please note that we will be able to give you a special discount of 3% if you place the order with us before March 31.
12. We suggest in your interest that you place an order earlier, as it is very likely that the price will rise soon.
13. Please quote us your lowest price, stating the terms of payment and packing condition.
14. We are sorry to say that as your price is too high. We will have to make our purchase elsewhere.
15. We thank you for your offer by fax of September 5 for 6,000 pieces of the captioned goods.
16. Please reply as soon as possible, stating the earliest date of shipment and the terms of payment.
17. If a substantial reduction can be expected, we could reconsider your revised quotation.
18. It is hoped that you would seriously take it into consideration and let us have your reply very soon.
19. We will send you a firm offer with shipment available in the early May if your order reaches us before March 10.
20. We find it very regrettable to point out that your prices are not very competitive and that there is no possibility of business.

Section C

clip 3-17

Listen to the audio clip 3-17 and type the following paragraphs sentence by sentence.

1. We have made a good selection of patterns and sent them to you today by post. Their fine quality, attractive designs and the reasonable prices at which we offer them will convince you that these materials are really of good value. There is a heavy demand for our supplies from house furnishers in various districts and regions, so we are finding it difficult to meet.
2. Being requested, we enclose our latest price list and catalogue of this month by air. A very full range of the blankets have been sent to you by sample post today, and we are confident that you will see the quality and the prices of our goods favorably compared with any others, for the same class of goods. We are continuously issuing new designs and we are delighted to submit further samples to you if there are orders from you in succession from now on.
3. It is our company's policy not to discount on our standard prices. However, we are glad to make an exception in this case as an introduction to our "Panda" Brand Color

TV Sets. Thus, we accept your counter-offer of a 5% discount based on a purchase of 1,500 sets by July 15.

Section D

Listen to the audio clip 3-18 and type the following letter sentence by sentence.

clip 3-18

A

Dear David,

Thank you for your offer dated April 20.

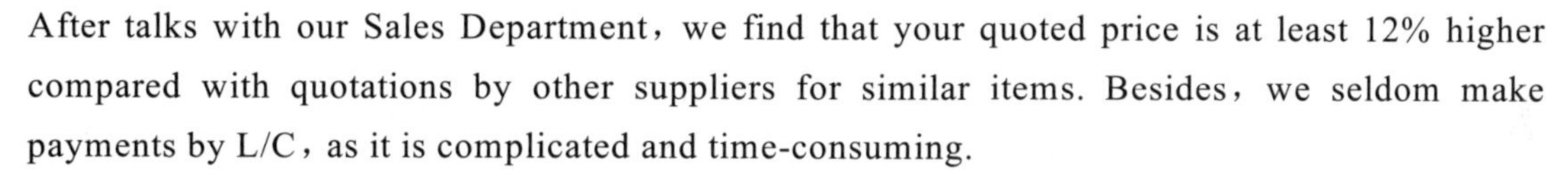
After talks with our Sales Department, we find that your quoted price is at least 12% higher compared with quotations by other suppliers for similar items. Besides, we seldom make payments by L/C, as it is complicated and time-consuming.

Please consider whether you can lower the price and change the payment terms to T/T, 30% on order, 50% before shipment and the remaining 20% on arrival. Otherwise, we see little prospect of concluding the deal with you.

Let us have your prompt feedbacks.

Regards,

Tommy

Listen to the audio clip 3-19 and type the following letter sentence by sentence.

clip 3-19

B

Dear John,

We appreciate your prompt feedbacks regarding our offer to you for bedroom LED lamps. But we're more than surprised to learn that you find our price too high and require a discount of 15%!

As you also know, there has been a significant rise in the prices of raw materials and in the costs of labor over the past few months. Taking the material of stainless steel as an example, the current price is 8.5% higher than that in July. This has put us under great pressure with respect to cost control. So we have to tell you with much regret that we cannot accept your counter-offer.

Considering your efforts in market development for our brand, we shall, however, try to give you a 5% discount if you would raise your order amount up to over USD 50,000. Please understand this is the best we can do.

Hope you will find it acceptable and will place an order soon.

Yours faithfully,

Sandy

clip 3-20

Listen to the audio clip 3-20 and type the following letter sentence by sentence.

 **C**

Dear Mike,

Thanks a lot for your mail dated June 18. We're glad that you accept our price and payment terms.

However, we're sorry to tell you that we cannot meet your requirement of delivery before mid-September. There has recently been a very big demand for cotton sheets and pillow cases. Actually, we already stop taking orders which require shipments in August and September. So for new orders, the earliest possible delivery date would be from October.

We understand that you want earlier delivery to catch the autumn and winter selling season. Considering our long-term business relationship, we will try our best to give you the goods as early as we can. But we would also suggest that you accept partial shipments.

Please kindly let us have your comments.

Regards,

Wendy Chen

Sales Manager

clip 3-21

Listen to the audio clip 3-21 and type the following letter sentence by sentence.

➢ **D**

Dear Susan,

Thanks very much for your prompt reply dated March 25.

We note that you found our price on the high side and required an 8% reduction. It is true that our price is a bit higher than those of other competitors. But our quality is beyond comparison. After review of our price calculations, we're sorry to tell you that we cannot agree to the 8% reduction you proposed.

We understand, however, that you will have some difficulties in selling the products, because the brand is still unknown to your customers. So, as a support to you, we've decided to give you for free 3000 pcs of Bluetooth headsets which you can use in your promotion activities. We believe that this will definitely help to boost the sales.

Please kindly let us have your comments.

Best regards,

Kevin Liu

Notes

reach an agreement		达成协议
lower price		降价
firm offer		实盘
T/T		电汇（Telegraphic Transfer）
marble floor tile		大理石地砖
on the high side		（价格）太高
furnisher	*n.*	家具商
in succession		接连地；连续地
time-consuming	*a.*	耗时的
prospect	*n.*	前景；可能性
conclude the deal		达成交易；成交
stainless steel		不锈钢
with respect to		关于；至于
take orders		接单
selling season		销售旺季
partial shipments		分批装运
Bluetooth headset		蓝牙耳机

3.4 Orders 订单

Part One: Warm-up Activities

In Part One, you will practice typing sentences, paragraphs and passages dealing with orders. Firstly you are supposed to read aloud and to identify the new words listed in Section A, and then try to practice typing the sentences, paragraphs and passages in Section B, C, D respectively under the teacher's guidance. While you are typing, please mark out the time you spend on each section and compare your results with your classmates.

Section A Vocabulary work

to our satisfaction		令我方满意
You may rest assured that…		贵方大可放心……
dispatch	*v.*	发货
New Orleans		新奥尔良（美国港口城市）
resume	*v.*	重新开始；继续

car coat		风衣
navy blue		海军蓝
order form		订单表
acknowledge receipt of		收到
allowance	*n.*	折扣
concession	*n.*	让步

Section B Sentence practice

1. We appreciate your cooperation.
2. Your order No.85 has been booked.
3. Thank you for your order No. 354GF.
4. Enclosed please find our Order No. B5421.
5. We hope to receive your order again.
6. We are pleased to receive your order and welcome you as one of our customers.
7. We will submit further orders, if this one is completed to our satisfaction.
8. We are pleased to confirm your order which we have accepted.
9. Your terms are satisfactory and we are pleased to place an order with you.
10. We have these clocks in stock and will be able to deliver them before March 6th as requested.

Section C Paragraph practice

1. We learn that an L/C covering the above-mentioned goods will be established immediately. You may rest assured that we will arrange for dispatch by the first available steamer with the least possible delay upon receipt of your L/C.
2. Thank you for your order No. C876 for 125 Do-It-Yourself Paint Machines. However, we are unable at this time to fulfill this order due to a fire in our manufacturing plant in New Orleans three days ago. We intend to resume production next week and expect to deliver your order early next month.
3. We are happy to enclose our trial order No. 8822, for 325 Burda Ladies' Car Coats, medium size, navy blue color; at USD 98.65 per piece, subject to six percent quantity discount. Please sign the duplicate of the enclosed order form and return it to us as your acknowledgment.

Section D Passage practice

➢ Passage 1

We acknowledge receipt of the captioned offer dated May 18. We, however, regret that we are unable to accept your offer as your prices are too high. We also have similar offers from other suppliers in the line. Their prices are 5% lower than yours. We appreciate the good quality of your products, but it does not justify such a large difference in price. We might do business with you if you could make us some allowance, say 5%, on your prices. Otherwise we have to decline your offer. We shall appreciate it very much if you will make a concession and email us your acceptance as soon as possible.

➢ Passage 2

We have received your letter of 21 May, from which we learn that the prices for captioned goods are found too high to be acceptable. We wish to inform you that considering the long-term business relationship between us, we have quoted you the most favorable prices. As a matter of fact, our price has been accepted by other buyers from your country where many transactions have been concluded. We believe our prices are fixed at a reasonable level. Therefore, we cannot satisfy your counter-offer for the time being. But if you could purchase 2,500 sets, we would grant you a 1% discount. We hope you will reconsider our offer and send us your order for our confirmation at your earliest convenience.

Part Two: Audio-typing

Now you are going to listen to the recording. Do not refer to your textbook while you are listening. Then you are supposed to listen to each section sentence by sentence once again and type what you hear at the same time.

Listen to the audio clip 3-22 and type the following short and simple sentences.

clip 3-22

1. Enclosed please find our Order No. 237 for four of the items.
2. We, therefore, hope you will make delivery at an early date.
3. Please send us your confirmation of sales in duplicate.
4. We thank you for your Order No. 6235 for delivery in February.
5. Our products are selling well in many areas.
6. Your order is booked and will be handled with great care.

7. Please let us know in case you are interested in any of the items.
8. We will, however, contact you by fax once supplying improves.
9. We appreciate your cooperation and look forward to receiving your further orders.
10. We regret that we cannot book the order at the prices we quoted six weeks ago.
11. We are looking forward to your acknowledgment.
12. Thank you for doing business with us.
13. We apologize for the delay and hope it will not cause you serious inconvenience.
14. We are sure you will be pleased with this new line of wrist watches.
15. We anticipate your good news next time.

Section B

clip 3-23

Listen to the audio clip 3-23 and type the following longer and more difficult sentences.

1. Our usual terms of payment are cash against documents and we hope they will be acceptable to you.
2. We thank you for your letter and are glad to inform you that your samples are satisfactory.
3. We require payment to be made by a confirmed and irrevocable letter of credit payable by draft at sight upon presentation of shipping documents.
4. It is understood that a letter of credit in our favor covering the above-mentioned goods will be established immediately.
5. We hope that our handling of this first order of yours will lead to further business between us and mark the beginning of a happy working relationship.
6. You may rest assured that we shall effect shipment with the least possible delay upon receipt of the credit.
7. We were compelled to adjust our prices in order to cover at least part of this increase.
8. We are in receipt of your letter of March 18 and are pleased to place an order with you for the following goods.
9. We cherish our cooperation and assure you that your order will receive our most careful attention.
10. Your order is receiving our immediate attention and we will keep you informed of the progress.
11. We are sorry to say that we must turn down your order as we have full order books at present and cannot give a definite date for delivery.
12. We are sorry that your order of Model 1 is out of stock now. Please select a suitable substitute from the enclosed catalogue.

13. You may rest assured that as soon as we are able to accept new orders, we shall give priority or preference to yours.
14. If the quality of your machine is satisfactory and your prices are right, we expect to place regular orders for fairly large numbers.
15. Please let us know the color assortment at once and open the covering L/C in our favor according to the terms contracted.
16. As stated in your quotation of April 8, we may expect immediate shipment from stock.
17. We would appreciate delivery within one month and look forward to your acknowledgment.
18. We are pleased to inform you that your order No. G229 is being processed and will be dispatched by airfreight to Naples on July 2.
19. We are sorry that we cannot supply your order on the credit terms you requested in your fax of March 3.
20. We place this trial order on the clear understanding that delivery to our warehouse in Singapore has to take place before May 1.

Section C

Listen to the audio clip 3-24 and type the following paragraphs sentence by sentence.

clip 3-24

1. We have received your letter of September 20 together with an order for 1,000 sewing machines. Enclosed is our Sales Confirmation No. 346 in duplicate, one copy of which please sign and return to us for our file.
2. As soon as we receive your reply in the affirmative, we shall confirm supply of the prints at the prices stated in your letter and arrange for dispatch by the first available steamer upon receipt of your L/C.
3. We wish to point out that the stipulations in the relevant credit should strictly conform to the terms stated in our Sales Confirmation in order to avoid subsequent amendments.
4. We find both quality and price satisfactory and are pleased to give you an order for the following items on the understanding that they will be supplied from current stock at the price named.

clip 3-25

Listen to the audio clip 3-25 and type the following letter sentence by sentence.

Dear Sirs,

We thank you for your quotation of July 3 for the supply of vacuum bottles and find your terms acceptable. We are pleased to enclose our order, No. 993 for 1,500 unbreakable stainless-steel vacuum bottles at USD19.65 per bottle.

We would appreciate delivery within one month and look forward to your acknowledgment.

Yours sincerely,

Susan

clip 3-26

Listen to the audio clip 3-26 and type the following letter sentence by sentence.

Dear Sir,

Thanks for your consideration for our counter-offer and your prompt reply.

I am pleased to tell you that in view of the great demand for the captioned goods, although the prices are still quite high, we would like to place an order for 2,500 sets of "Panda" Brand Color TV Sets at USD 445 per set CIF Sydney.

Please send us your Sales Confirmation for our counter-signature. If everything is in order, we will open the covering L/C on time.

We are looking forward to your prompt attention to this order.

Yours truly,

Mike

clip 3-27

Listen to the audio clip 3-27 and type the following letter sentence by sentence.

Dear Sir,

Thanks for your prompt reply regarding the captioned order and we take the pleasure in confirming the acceptance of your order.

We wish to inform you that due to the competitive price and fashionable design, our TV sets sell well in the world, which will help you increase your market share steadily. As to the shipment, we will provide punctual shipment as what is said in the contract.

Enclosed is our Sales Contract No. 792. Please counter-sign and return it for our file.

We are looking forward to your covering L/C at an early date.
Yours faithfully,
Linda

Listen to the audio clip 3-28 and type the following letter sentence by sentence.

clip 3-28

 **D**

Dear Sirs,
We thank you for your order of September 10 for brown serge, but regret being unable to entertain it because the original price in our offer of August 15th has risen considerably. We ourselves even could hardly replenish our stock at the price quoted previously. It is hoped that you will understand our situation.
We may inform you that the ruling price for the article on your order is now at USD1.05 per yard FOB Shanghai. However, in order to finalize this first transaction between us, we are prepared to allow you a 2% discount on the above rate.
We sincerely recommend you to accept our proposal as our stocks are getting lower day by day, and we are afraid we shall be unable to meet your requirements if you fail to let us have your confirmation by return.
We look forward with pleasure to receiving your reply.
Yours faithfully,
Smith

Notes

anticipate	*v.*	期待
cash against documents		凭单付现
payable by draft at sight		凭即期汇票兑付;即期支付
upon presentation of		出示……;凭……
in one's favor		以……方为受益人
effect shipment		实施装船
compel	*v.*	迫使;促使
out of stock		无货;缺货
substitute	*n.*	替代品
Naples		那不勒斯(意大利西南部港口城市)
airfreight	*n.*	空运
warehouse	*n.*	仓库
Enclosed is…		随附……

Sales Confirmation		成交确认书
in duplicate		一式两份
for our file		供我方存档
affirmative	*a.*	肯定的
steamer	*n.*	轮船
stipulation	*n.*	规定;条款
conform to		符合;一致
subsequent amendment		后续的修改
current stock		现有库存;现货
at the price named		按所示的价格
vacuum bottle		热水瓶;保温瓶
unbreakable	a.	不易破损的
in view of		鉴于……;考虑到……
counter-signature	*n.*	会签;回签
the covering L/C		相关信用证
market share		市场份额
serge	*n.*	哔叽
entertain	*v.*	接受
ruling price		市价;时价
finalize	*v.*	完成;使结束
by return		回信;立即回复

Chapter 4
Trade Procedure
贸易流程

4.1 Business Contracts 商务合同

Part One: Warm-up Activities

In Part One, you will practice typing sentences, paragraphs and passages dealing with business contracts. Firstly you are supposed to read aloud and to identify the new words listed in Section A, and then try to practice typing the sentences, paragraphs and passages in Section B, C, D respectively under the teacher's guidance. While you are typing, please mark out the time you spend on each section and compare your results with your classmates.

Section A Vocabulary work

sales contract		销售合同
allied	*a.*	联合的
quantity	*n.*	数量
destination	*n.*	目的地
Rotterdam		鹿特丹(荷兰西南部港口城市)
under-mentioned	*a.*	下述的
negotiation	*n.*	商议;谈判
in accordance with		与……一致;依照
shipping mark		装运标记
packing clause		包装条款
be entitled to		有权利……

add this provision		加列这项条款
seaworthy	*a.*	适于航海的
reinforced cardboard box		加固的纸板箱
invoice value		发票价值
insurance clause		保险条款

Section B Sentence practice

1. Let's check the items in the sales contract.
2. Sellers: Hubei Provincial Light Industrial Products Imp. / Exp. Corp.
3. Buyers: Allied Trading Company, Lagos, Nigeria.
4. We've finally come to an agreement.
5. All the content is written both in Chinese and in English.
6. Please check the name of the commodity, specifications, quantity, unit price and the total amount.
7. Here's the contract we drafted according to the outcome of our negotiations.
8. Time of Shipment: on 15 May, 2021.
9. Port of Shipment: Shanghai, China.
10. Port of Destination: Rotterdam.

Section C Paragraph practice

1. This Contract is made by and between the Buyers and the Sellers, whereby the Buyers agree to buy and the Sellers agree to sell the under-mentioned commodity according to the terms and conditions stipulated below.
2. Our commodities have to go through strict inspection in accordance with the stipulations in the contract before shipment. They can be exported only when they're up to the required standard.
3. Terms of Payment: By irrevocable letter of credit payable by draft at sight. The L/C should reach the Seller 30 days before the time of shipment and remain valid for negotiation in China until the 15th day after the date of shipment.

Section D Passage practice

➤ Passage 1

A: I forgot to mention the shipping marks when we discussed the packing clause.

B: For that we can add a sentence like "the shipping marks will be provided according to the buyer's samples" in the contract. If the patterns of those shipping marks are on

hand now, you can mail them to us after you return to the United States.

A: Thanks a lot. Another thing on my mind is that I think we should add this provision, "If one side fails to honor this contract, the other side is entitled to cancel this contract." What do you think?

B: I think that's fair. We certainly should include this provision in our contract.

➢ Passage 2

Buyers: the General Trading Company, Japan

Sellers: China National Native Product Imp./Exp. Co. Wuhan Branch

Commodity: Chinese Dinner Sets

Specifications: Art. No. 5674

Quantity: 600 sets

Unit Price: at US $45 per set CIF Tokyo

Packing: in sea worthy reinforced cardboard boxes

Insurance: to be effected by the sellers for 110% of the invoice value against All Risks and Breakage as per China Insurance Clauses

Time of Shipment: in May, 2021

Port of Shipment: Wuhan, China

Port of Destination: Tokyo, Japan

Shipping Marks: at seller's option

Terms of Payment: by irrevocable sight L/C

Done and signed in Wuhan on the third day of November, 2021

Part Two: Audio-typing

Now you are going to listen to the recording. Do not refer to your textbook while you are listening. Then you are supposed to listen to each section sentence by sentence once again and type what you hear at the same time.

Section A

Listen to the audio clip 4-1 and type the following short and simple sentences.

clip 4-1

1. Now, let's sign the contract.
2. When will the contract be ready?
3. First of all, let's talk about the format of our sales contract.
4. As long as you've got an English version, I have no objection.

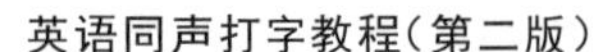

5. We can't accept any delay.
6. Our price is determined on the basis of CIF.
7. You can contact us any time tomorrow before 5 p.m.
8. At 3 o'clock in the afternoon, we'll sign the contract.
9. I forgot to mention the materials we should use.
10. We should delete this provision.
11. We should add one more clause to the contract.
12. The copies of the sales contract are all here now.
13. We should be more direct when writing this contract.
14. We really should honor the contract without any excuse.
15. Please take one more look before you sign.
16. We have a few remarks to make on the contractual obligations of both parties.
17. The force majeure clause is missing in the contract.
18. For the convenience of our both parties, we have checked the contract once more.
19. That was a big contract. The negotiation was hard and lasting.
20. In principle we agree with most of the clauses referring to the general conditions.

Section B

clip 4-2

Listen to the audio clip 4-2 and type the following longer and more difficult sentences.

1. Each of us has two formal copies of the contract: one in Chinese and one in English.
2. I want to point out here that shipment should be effected and completed before the end of October.
3. We certainly should include this provision in our contract and now the clauses of the contract are good enough.
4. I think that's a good idea and it will help clarify some important items that we may have overlooked.
5. The contract is ready now and you may go over the contract and see if everything is in order.
6. This is not important in terms of a long-term agreement and we could meet every six months or so to set the price.
7. We'd like to add the condition that the L/C shall be valid until the 15th day after shipment.
8. Let's go over all the terms and conditions of the transaction to see if we agree on all the particulars.
9. I think the situation is right for us to talk about a long-term agreement.

10. If it' s agreeable to you, Miss Susan could prepare the agreement, and it would be ready for our signatures tomorrow afternoon.
11. I think this clause suits us well, but the time of payment should be prolonged three or four months.
12. One of our principles is that contracts are honored and commercial integrity is maintained.
13. It contains basically all we have agreed upon during our negotiation.
14. Thanks to your open policy, without which our cooperation would be impossible.
15. The Contract No. should be No. AB106 instead of AB601.
16. We wish to make it clear hereby that once a contract is signed, it has legal effect, so no party who has signed a contract has the right to break it.
17. Since all the contracts are to be executed according to the principle of "first come, first served", we could recommend you to sign the contract and return it to us without further delay.
18. The new packaging of this article is exquisitely designed and we are confident that it will appeal strongly to consumers.
19. Buyer shall furnish Seller with necessary instructions for make-up, description of origin, packing, marking and/or other arrangements in time for preparation of shipment of the goods respectively.
20. The Buyers are requested to sign and return one original copy to the Seller for file immediately upon receipt of this Sales Confirmation.

Section C

Listen to the audio clip 4-3 and type the following paragraphs sentence by sentence.

clip 4-3

1. It contains basically all we have agreed upon during our negotiation. The last point is that the inspection should be carried out by Wuhan Commodities Inspection Bureau, whose decision is the final basis and binding on both parties.
2. If we weren't able to buy as many as we wrote down in the contract for one reason or another in one year, or you couldn't supply us with as many as in the contract, what would be done about the difference?
3. Let's add on one more sentence here, "In the event the breaching party has paid part of its investment in accordance with the stipulation of the contract, the joint venture shall clear up its investment."
4. We enclose S/C No. 457 in duplicate, of which please countersign and return one copy to us for our file. Shall the Buyers fail to do so within 10 days after arrival of this Sales

Confirmation at the buyers' end, it shall be considered that the Buyers have accepted all the terms and conditions set forth in this Sales Confirmation.

5. A usual trade margin of 5% plus or minus of the quantities confirmed shall be allowed. When shipment is spread over two or more periods, the above-mentioned trade margin of plus or minus 5% shall, when necessary, be applicable to the quantity designated by the Buyers to be shipped each period.

Section D

clip 4-4

Listen to the audio clip 4-4 and type the following contract sentence by sentence.

➢ **A**

SALES CONTRACT

No. 290

Date: August 20, 2020

Sellers: China National Light Industrial Products Imp./Exp. Co.

Buyers: Cathy Company, HK

This CONTRACT is made by and between the Buyers and the Sellers, whereby the Buyers agree to buy and the Sellers agree to sell the undermentioned goods on the terms and conditions stipulated below:

Commodity: Working Gloves

Specifications: Type No. 572

Quantity: 5,000 dozen

Unit Price: HK$ 27.00 per dozen CIF Hong Kong

Total Value: HK$ 135,000.00 (Say Hong Kong Dollars One Hundred Thirty-Five Thousand Only)

Packing: In boxes of a dozen each, 10 boxes to a carton

Insurance: To be covered by the Sellers against All Risks and War Risk for 110% of the invoice value

Terms of Payment: By confirmed irrevocable L/C payable by draft at sight, to reach the Sellers 30 days before shipment and remain valid for negotiation in China till the 15th day after the latest date of shipment as specified in the covering L/C

Done and signed in Beijing on this 5th day of November, 2020.

Listen to the audio clip 4-5 and type the following contract sentence by sentence.

clip 4-5

 **B**

CONTRACT

NO. 94HTS-457

Buyers: The Eastern Traders of New York, USA
Sellers: Great Wall Imp. & Exp. Co., Beijing
Commodity: Chinese Straw Hats
Specification: Art. No. 8774
Quantity: 1,000 dozen
Unit Price: At US $ 7.00 per dozen CIF New York
Total Value: US $ 7,000.00
Packing: 4 dozen per carton
Shipping Mark: At Sellers' option
Insurance: To be effected by the Sellers for 110% of the invoice value against All Risks and War Risk as per the China Insurance Clauses
Time of Shipment: During November, 2021 with partial shipment and transshipment allowed
Port of shipment: China Port(s)
Port of Destination: New York, USA
Terms of Payment: By a sight irrevocable L/C available by draft at sight to reach the Sellers one month prior to the time of shipment

Listen to the audio clip 4-6 and type the following agreement sentence by sentence.

clip 4-6

 **C**

AGREEMENT FOR EXCHANGE AND COLLABORATION BETWEEN THE STUDENTS OF HUBEI INSTITUTE OF EDUCATION IN CHINA AND SHERIDAN INSTITUTE OF TECHNOLOGY OF CANADA

The purpose of this agreement is to facilitate the exchange of students between Hubei Institute of Education in China and Sheridan Institute of Technology of Canada. Both institutes agree that collaboration and exchange should be conducted among students in academic activities as well as cultural exchanges.

It is further agreed that students of the above-mentioned institutes participate in all the school academic activities and share cultural experiences and other materials which are deemed to be of mutual institutional benefit.

The collaborative activities to be undertaken are to take place in accordance with existing regulations at both institutes and within the constraints imposed by the fiscal and human resources available.

Final approval of the collaborative programs to be developed will rest with the dean's corresponding authorities of both institutes.

This agreement shall exist for a period of three years, after which it may be extended, amended or discontinued at the initiative of either Hubei Institute of Education in China or Sheridan Institute of Technology of Canada.

Signature	Signature
______________________	______________________
President, Hubei Institute of Education	President, Sheridan Institute of Technology

clip 4-7

Listen to the audio clip 4-7 and type the following contract sentence by sentence.

➤ **D**

CONTRACT

No. 54-76
Sellers: Guangzhou Export Corporation
Buyers: Far East Trading Company Ltd.

This contract is made by and between the Buyers and the Sellers, whereby the Buyers agree to buy and the Sellers agree to sell the undermentioned commodity according to the terms and conditions stated below:

Commodity: Ladies' Pajamas
Specifications: Art. No. 801
Sizes: S/3, M/6, and L/3 per dozen
Colors: Pink, blue and yellow, equally assorted
Quantity: 3,000 dozen /sets
Unit Price: £26.00 per dozen/set CIF London
Total Value: £78,000.00
Packing: In cartons

Shipping Mark: At buyer's option
Insurance: Covering marine All Risks and War Risk for 100% of CIF value plus 10%
Time of Shipment: In December 2021, with transshipment at London
Port of Shipment: Guangzhou
Port of Destination: London
Terms of Payment: By irrevocable sight L/C to be opened in seller's favor 30 days before the time of shipment
Done and signed in Guangzhou on this 20th day of September 2021.

Listen to the audio clip 4-8 and type the following agreement sentence by sentence.

clip 4-8

 **E**

This agreement constitutes the entire and only agreement between the parties hereto and supersedes all previous negotiations, and agreements relating to the sales of products and shall not be modified or changed in any manner except mutual consent on writing of the subsequent date signed by a duly authorized officer or representative of each of the parties hereto.

If the buyer fails to furnish, promptly upon request, any details and instructions necessary to enable the seller to perform the contract in accordance with its terms, the seller shall be entitled, at its option, and in addition to all other rights, to cancel such portion of the contract as may remain unexecuted, or to make shipment in accordance with the details and instructions which the buyer may have furnished for previous shipments on account of the same or a previous contract. The buyer shall not, however, be entitled to change or modify, except with the consent of the seller, any details or instructions comprised in the contract itself.

Notes

force majeure clause		不可抗力条款
particular	*n.*	细节
signature	*n.*	签名；署名
prolong	*v.*	延长；拖延
integrity	*n.*	诚实；完整性
legal effect		法律效力
execute	*v.*	执行；使生效
exquisitely	*ad.*	高雅地；精致地
make-up	*n.*	外观
Commodities Inspection Bureau		商品检验局
binding	*a.*	具有约束力的

breaching party		违约方
clear up		结清
the joint venture		合资企业
S/C(= Sales Contract)		合同
countersign	*v.*	连署
margin	*n.*	差数
5% plus or minus		5%的增减
carton	*n.*	纸板箱
transshipment	*n.*	转载;转运
prior to		在前;居先
facilitate	*v.*	(不以人作主语的)推动;促进
academic	*a.*	学院的;理论的
deem	*v.*	认为;相信
regulation	*n.*	规则
constraint	*n.*	约束
amend	*v.*	修正
initiative	*n.*	主动;率先
pajamas	*n.*	睡衣
assorted	*a.*	多样混合的
marine	*a.*	海运的
hereto	*ad.*	关于这个;到此为止
supersede	*v.*	代替;取代
modify	*v.*	更改;修改
duly authorized		正式授权的
option	*n.*	选择权
comprise	*v.*	包含

4.2 Payment 支付

Part One: Warm-up Activities

In Part One, you will practice typing sentences, paragraphs and passages dealing with payment. Firstly you are supposed to read aloud and to identify the new words listed in Section A, and then try to practice typing the sentences, paragraphs and passages in Section B, C, D respectively under the teacher's guidance. While you are typing, please mark out the time you spend on each section and compare your results with your classmates.

Section A Vocabulary work

terms of payment		付款方式
D/P(Document against Payment)		付款交单
reimbursement	*n.*	偿还;偿付
draw	*v.*	出票;签出汇票
draft	*n.*	汇票
D/A(Documents against Acceptance)		承兑交单
expedite	*v.*	加快;促进
confirmed	*a.*	保兑的
sight	*n.*	见票
documentary draft		跟单汇票
presentation	*n.*	提示
maturity	*n.*	到期
shipping document		货运单据
bill of lading		海运提单
in triplicate		一式三份
insurance policy		保险单
shipping advice		装运通知

Section B Sentence practice

1. Now the terms of payment have been settled.
2. Payment is to be made before the end of next month.
3. We only accept D/P with new customers like you.
4. In reimbursement of these extra expenses, please draw a draft on us.
5. With regard to contract No.111, we are agreeable to D/A payment terms.
6. Please expedite the L/C so that we may execute the order smoothly.
7. Please amend the amount of the L/C to read "2% more or less".
8. We have issued a confirmed irrevocable letter of credit covering our order No. 53.
9. We regret we cannot accept "cash against documents on arrival of goods at destination".
10. We enclose a check, valued US $10.000, in payment of the account for goods delivered in Nov.

Section C Paragraph practice

1. Payment by D/P after sight. The buyer shall duly accept the documentary draft at 60 days sight upon first presentation and make payment on its maturity. The shipping

documents are to be delivered against payment only.

2. Payment by D/A. The buyer shall duly accept the documentary draft drawn by the seller at 90 days sight upon first presentation and make payment on its maturity. The shipping documents are to be delivered against acceptance.
3. The buyer shall open an irrevocable L/C in favor of the seller before May 8th，2021. The said L/C shall be available by draft at sight for full invoice value and remain valid for negotiation in China for 15 days after shipment.

Section D Passage practice

Dear Sirs，

With reference to our faxes dated the 5th of February and 10th of March，requesting you to establish the L/C covering the above mentioned order，we regret having received no news from you up till now.

We wish to remind you that it was agreed，when placing the order，that you would establish the required L/C upon receipt of our confirmation. As goods have been ready for shipment for quite some time，it behaves you to take immediate action，particularly since we cannot think of any valid reason for further delay of opening the credit.

We look forward to receiving your favorable response soon.

Yours faithfully，

Dear Sirs，

We have instructed the Bank of Boston，Massachusetts，to open an irrevocable letter of credit for USD 28，000 in your favor，valid until 28th，April.

The documents required for negotiation are：

Commercial invoice in duplicate；

Bills of lading in triplicate；

Insurance policy in one original and three copies.

Please make sure that the shipment is effected within April，since prompt delivery is one of the important considerations in dealing with our market.

We are looking forward to your shipping advice.

Yours sincerely，

Part Two: Audio-typing

Now you are going to listen to the recording. Do not refer to your textbook while you are listening. Then you are supposed to listen to each section sentence by sentence once again and type what you hear at the same time.

Section A

Listen to the audio clip 4-9 and type the following short and simple sentences.

clip 4-9

1. Payment by L/C will tie up our funds.
2. Please extend the L/C by 30 days.
3. We can't accept payment on deferred terms.
4. How about 50% by L/C and the rest by D/P?
5. We enclose a check for US$6,000.
6. We duly received your draft for $500.
7. Please do your utmost to expedite L/C.
8. Why was our draft No. 719 dishonored?
9. We may have some difficulties making payment in Euro.
10. We look forward to receiving your L/C order No. 456.
11. It would help us greatly if you would accept D/A instead.
12. 10% of the total amount of the deal should be paid in advance.
13. The stipulations in the L/C are not in agreement with the contract.
14. D/P will be accepted if the amount involved is not up to US$1,000.
15. Your commission of US$2,000 was remitted to your Beijing Office.

Section B

Listen to the audio clip 4-10 and type the following longer and more difficult sentences.

clip 4-10

1. To our regret, we find three points in the L/C do not conform to the contract.
2. We will open an L/C if you promise to effect shipment one month earlier.
3. We may accept deferred payment if the quantity is over 10,000 pieces.
4. As requested, we have extended the expiry date of the L/C to the end of this year.
5. We're still looking forward to your urgent reply to our settlement sheet for 2021 final payment.
6. If these terms are agreeable to you, please amend the relative L/C to allow partial

shipment.

7. As agreed, the terms of payment for the above orders are letter of credit at 60 days sight.
8. We confirm that you have paid on shipment 11/223, and want to check whether you have received our invoice.
9. As to payment, we are agreeable to draw on you at 30 days sight upon presentation and make payment at maturity.
10. The buyer should pay 100% of sales in advance by D/D to reach the seller not later than September 30.
11. We should be obliged for your immediate amendment of the L/C to enable us to make shipment in time.
12. The buyer shall pay the total value to the seller in advance by T/T (M/T or D/D) not later than June 30, 2021.
13. We must point out that unless your L/C reaches us by the end of this month, we shall not be able to effect shipment within the stipulated time limit.
14. Please send your remittance to our account No. 6738483 with the Bank of China Wuhan Branch in favor of SANYI Company.
15. We have the pleasure in advising you that L/C No. CN027 for the amount of US$ 50, 000 has been established in your favor through the commercial bank.
16. Owing to some delay on the part of our suppliers, please kindly extend the shipping date and credit validity for one month respectively.
17. In view of the amount of this transaction being very small, we are prepared to accept payment by D/P at sight for the value of the goods shipped.
18. The expiration date of the credit being March 6th, we request that you will arrange with your banker to extend it up to April 1st, amending the said credit.
19. In order to save a lot of expenses on opening the L/C, we will remit you the full amount by T/T when the goods ordered by us are ready for shipment and the freight space is booked.
20. Our terms of payment are by confirmed irrevocable L/C available by draft at sight, reaching us one month ahead of shipment, remaining valid for another 21 days after the time of shipment.

Section C

clip 4-11

Listen to the audio clip 4-11 and type the following paragraphs sentence by sentence.

1. It will interest you to know that as a special sign of encouragement, we shall consider

accepting payment by D/P during this promotion stage. We trust this will greatly facilitate your efforts in sales, and we await your favorable reply.

2. We wish to draw your attention to the fact that the date of delivery of your order No. 715 is approaching, but we have not received the covering L/C up to date. Please do your utmost to expedite it to reach here before Dec. 20, so that shipment may be effected without delay.
3. We received your final invoice of shipment 08/219. on the basis of which the balance amounts should be adjusted as US $ 9, 202.68. Therefore, your final settlement according to the final invoice should be US$ 7,621.23.

Section D

Listen to the audio clip 4-12 and type the following letter sentence by sentence.

clip 4-12

Dear Sirs,

We wish to place with you an order for 100 leather handbags.

For this particular order we would like to pay by D/P at sight. Involved about USD 2,000, this order is comparatively a little one. It goes without saying that we very appreciate the support you have extended to us in the past. If you can do us a special favor this time, please send us your contract, upon receipt of which we will establish the draft of D/P.

Yours faithfully,

Listen to the audio clip 4-13 and type the following letter sentence by sentence.

clip 4-13

Dear Sirs,

We are in receipt of your letter of August 22.

Having studied your suggestion for payment by D/P at sight for USD 2,000, it is quite difficult for us to accept that. As our usual practice goes, we require payment by confirmed and irrevocable letter of credit. That is to say, at present, we can't accept D/P terms in all transactions with our customers abroad.

I regret to say that we must adhere to our usual practice and hope that this will not affect our future business relations.

Yours sincerely,

clip 4-14

Listen to the audio clip 4-14 and type the following letter sentence by sentence.

Dear Sirs,

Thank you for your letter of 4th of June, which arrived this morning.

We are pleased that you have been able to ship our order in good time but we are surprised that you still demand payment against documents. After long years of satisfactory trading we feel that we are entitled to easier terms. Most of our suppliers are drawing on us at 60 days, documents against acceptance, and we should be grateful if you could grant us the same terms.

We are looking forward to your favorable reply.

Yours faithfully,

clip 4-15

Listen to the audio clip 4-15 and type the following letter sentence by sentence.

Dear Sirs,

We have received your letter in which you ask for an easier term of payment.

In consideration of the very pleasant business relationship we have had with your firm for more than 15 years, we have decided to agree to your suggestion. We shall, therefore, in future draw on you at 60 days, documents against acceptance, and trust that this term will suit your requirements.

We hope that our concession will result in a considerable increase of your orders and assure you that we shall always endeavor to execute them to your complete satisfaction.

Yours faithfully,

clip 4-16

Listen to the audio clip 4-16 and type the following letter sentence by sentence.

Dear Sirs,

We have received your L/C No. 378265, but regret to find that there are certain points which are not in accordance with the terms of our sales confirmation No. 2268.

1. Our S/C No. 2268 emphasized that your letter of credit shall allow transshipment, but your credit states: "transshipment prohibited".

2. Your L/C calls for shipment in two equal monthly lots during June/July, 2021, whereas it is explicitly stipulated in the S/C that shipment is to be made in a single lot not later than July 31, 2021.

3. There is no word "about" before the quantity and amount in the L/C, although the word is

clearly used before them in our S/C.

4. Your L/C stipulates for the goods in an assortment of Types A, B, C: 20%, 40%, 40% respectively. But the assortment contracted for is: Types A 30%, B 30% and C 40%.

You are requested to make the necessary amendments, as the shipment is conditional upon the conformity of the L/C with our sales confirmation.

We are awaiting your bank's amendment of the L/C. Thank you for your help.

Allen

Notes

deferred	*a.*	延期的
dishonor	*v.*	拒付
remit	*v.*	汇款
settlement sheet		决算表
M/T		信汇
D/D		票汇
validity	*n.*	有效期
involve	*v.*	牵涉;涉及
extend	*v.*	给予
adhere	*v.*	坚持;遵守
conformity	*n.*	一致
assortment	*n.*	分类;种类

4.3 Delivery 发货

Part One: Warm-up Activities

In Part One, you will practice typing sentences, paragraphs and passages dealing with delivery. Firstly you are supposed to read aloud and to identify the new words listed in Section A, and then try to practice typing the sentences, paragraphs and passages in Section B, C, D respectively under the teacher's guidance. While you are typing, please mark out the time you spend on each section and compare your results with your classmates.

Section A Vocabulary work

advance	*v.*	将……提前;使……前进
lot	*n.*	(一)批;全部

hasten	*v.*	加速;赶快
extension	*n.*	延期
S.S (steamship)	*n.*	轮船
consignment	*n.*	托付货物;托卖货物
punctual	*a.*	按时的;准时的
lodge a claim		索赔
backlog	*n.*	积压
China Post Air Mail		中国邮政(航空)小包
asap (as soon as possible)		尽快
track	*v.*	查询;追踪
enquiry	*n.*	咨询
free shipping		免运费
UPS		(美国)联合包裹运送服务公司(United Parcel Service, Inc.)
DHL		敦豪速递公司(Dalsey, Hillblom and Lynn,一家国际快递公司)

Section B Sentence practice

1. May I have your delivery date?
2. Could you possibly advance shipment?
3. You can deliver the goods partially.
4. We insist on your prompt delivery.
5. When is the earliest shipment we can expect?
6. We will deliver the whole lot within 3 days.
7. Orders are delivered within 3-60 working days.
8. The shipping methods are set up by the suppliers.
9. The goods will arrive on time next week.
10. You must have the goods delivered before November.

Section C Paragraph practice

1. Please do your very best to hasten shipment. We hope that by the time you receive this message, you will have the goods ready for shipment. Any further extension will not be considered.
2. We take this opportunity to inform you that we have this day shipped the goods on board S.S "East Wind" which sails for your port tomorrow. Enclosed please find one set of the shipping documents covering this consignment.

3. As our company is in urgent need of the goods, we would like to emphasize again the importance of the punctual shipment. In case you should fail to effect shipment by the end of this month, we would have to lodge a claim against you for the loss.

Section D Passage practice

➢ Passage 1

Seller: My dear friend, there is a backlog of orders for China Post Air Mail to ship. I don't know when your packet can be shipped. How about changing a logistics company?
Buyer: OK, no problem. I only wish I can get my packet asap.
Seller: Dear friend, I have changed a logistics company. Now you can track your packet by this e-packet Number 12875.

➢ Passage 2

Dear customer,
Thank you for your enquiry about the shipping and delivery we offered. Our items are free shipping to most countries by China Post Registered Air Mail. It usually takes 15-21 days to reach your country. If you need the items urgently, we also offer the following express shipping options: UPS, FedEx, DHL, EMS. But you have to pay the extra freight according to the real cost.
Best regards,

Part Two: Audio-typing

Now you are going to listen to the recording. Do not refer to your textbook while you are listening. Then you are supposed to listen to each section sentence by sentence once again and type what you hear at the same time.

Listen to the audio clip 4-17 and type the following short and simple sentences.

clip 4-17

1. How long was the delay?
2. When can I get my package?
3. Have you shipped out?
4. How soon can we have the goods?
5. I can't track my package.
6. Please arrange immediate shipment.

7. We'll do our best to effect shipment.
8. We trust you will deliver the goods in time.
9. Your item will be sent out in 3 business days.
10. The delivery will be delayed by 4 days.
11. The delivery time is March or April at our option.
12. The shipment per S.S "Mercury" has gone forward.
13. You can check shipping time on the product detail page.
14. We apologize that the shipping is a little slower than usual.
15. We have also extended the time for you to confirm delivery.

Section B

clip 4-18

Listen to the audio clip 4-18 and type the following longer and more difficult sentences.

1. You can check the shipping information 2-3 days later on the Internet.
2. We hope you will let us have your shipping advice without further delay.
3. You may also choose alternative shipping methods to lower the shipping cost.
4. Currently suppliers can choose from the following express shipping options.
5. This item has been shipped out to your country and I suggest you wait for a few days.
6. Once the payment is confirmed，we'll process the order and ship it out as soon as possible.
7. Your items are sent out by China Post Air Mail and they will take out 15 working days to arrive.
8. Generally speaking，your package needs 2-5 days to pack up，and 3-8 weeks to arrive in your country.
9. Please ship the first lot under Contract No.123 by S.S "Long March" scheduled to sail on or about December 2.
10. Our after-sales service will keep tracking it and send message to you when there is any delay in shipping.
11. We trust you will ship the order within the stipulated time as any delay would cause us no little inconvenience and financial loss.
12. We shall appreciate it if you will effect shipment as soon as possible，thus enabling our buyers to catch the brisk demand at the start of the season.
13. We'd like to draw your attention to the fact that up to the present moment，we haven't received your shipping instructions referring to Order No.890003.
14. In compliance with the terms of contract，we forward you by air a full set of non-negotiable documents immediately after the goods were shipped.

15. Please try your utmost to deliver our goods by S.S "Peace" which is due to arrive at Hamburg on May 5, and confirm by return that the goods will be ready in time.
16. As the suppliers cannot get the entire quantity ready at the same time, it is necessary for the contract stipulations to be so expressed as to allow partial shipment.
17. Our government has recently put an embargo on the export of various medicines to your area and we have to obtain a special license to execute your order No.3982.
18. We regret our inability to comply with your request for shipping the goods in early April, because the direct steamer sailing for Melbourne calls at our port only around 20th every month.
19. It is imperative that you notify us immediately of the earliest possible date of shipment for our consideration without prejudice to our right of cancelling the order or lodging claims for losses.
20. We are pleased to inform you that the goods under your order No.7582 were shipped by the "Red Star" on Nov. 30 and the relevant shipping documents had been dispatched to you by air.

Section C

Listen to the audio clip 4-19 and type the following paragraphs sentence by sentence.

clip 4-19

1. We'd like to draw your attention to the fact that up to the present moment, we haven't received your shipping documents. It's explicitly stipulated in the sales confirmation that shipment must be effected by the end of this month.
2. I am so sorry for the delivery delay, as we had a 5-day off in this week. After holiday, we must deal with large amounts of order, so that lots of parcels had not been shipped out in time. We are very sorry for the delay and hope you can understand us.
3. Because of the fragile nature of our goods, we provide home delivery service to our customers within Wuhan. Our delivery day is Friday, which means your items will arrive this coming Friday, July 16. If this is unsatisfactory, please call us so that we can arrange alternative delivery date.

Section D

Listen to the audio clip 4-20 and type the following letter sentence by sentence.

clip 4-20

A

Dear buyer,

Thanks for your order with us, but we are sorry to tell you that due to peak season these days, the shipping to your country was delayed.

We will keep tracking the shipping status and keep you posted of any update.

Sorry for the inconvenience caused, and we will give you 5% off to your next order for your great understanding.

Best Regards,

Lauren

clip 4-21

Listen to the audio clip 4-21 and type the following letter sentence by sentence.

 **B**

Dear customer,

Thank you for shopping.

We have shipped out your order on July 16th by China Post Air Mail. You may login www.17track.net to track your item 3 days later with the tracking number 7865455. It will take 5-20 workdays to reach your country. Hope you'll be satisfied with our services as well as the product and we look forward to your feedback soon. If you have any questions, please feel free to contact us at any time.

Best regards,

Nancy

clip 4-22

Listen to the audio clip 4-22 and type the following letter sentence by sentence.

➢ **C**

Dear Sirs,

This refers to our previous letter in which we asked about the delivery of 200 dozen cashmere sweaters under our S/C No. 889.

It's a great pity that we haven't received any news concerning these goods up till now. Now the cold season is approaching and our customers are in urgent need of the goods. We have assured them of the timely delivery of the goods, so we hope that you will do your utmost to arrange for the prompt shipment as soon as possible. In case you fail to effect delivery before the end of the month which is originally agreed, we'll have to lodge a claim against you for the losses incurred thereby.

We are waiting for your reply.

Sincerely yours,

Listen to the audio clip 4-23 and type the following letter sentence by sentence.

clip 4-23

 **D**

Dear Sirs,
We wish to call your attention to the fact that up to the present moment no news has come from you about the shipment under the captioned contract.
As you have been informed in one of our previous letters, the users are in urgent need of goods contracted and are in fact pressing us for assurance of early delivery.
Under the circumstances, it is obviously impossible for us to again extend L/C, which expires on September 11, and we feel it our duty to remind you of this matter again.
As your prompt attention to shipment is most desirable to all parties concerned, we hope you will let us have your shipping advice without further delay.
Yours truly,

Listen to the audio clip 4-24 and type the following letter sentence by sentence.

clip 4-24

 **E**

Dear Sirs,
Thanks for your Email of May 8 and 10 together with your order No. 5678. We enclose our S/C No. 6632 in duplicate, a copy of which is to be returned to us after being countersigned by you.
We note that you require direct shipment. As there are only limited number of direct sailings from here to the port of destination, delivery will possibly be delayed. It is desirable to have the goods transshipped via Hong Kong or elsewhere. For this reason, we would request you to allow transshipment in the relevant L/C so as to avoid subsequent amendments.
Shipment will be effected within 15 days after the receipt of your L/C.
Yours faithfully,

Notes

alternative	*a.*	选择性的;代替的
express	*n.*	快递
process	*v.*	处理
brisk	*a.*	火热的
in compliance with		按照
non-negotiable	*a.*	不可转让的
embargo	*n.*	贸易禁运
imperative	*a.*	必要的

without prejudice to		无害于;不影响
explicitly	*ad.*	清楚地;明白地
fragile	*a.*	易损的
peak season		旺季
login	*v.*	登录
incur	*v.*	遭受

4.4 After-sales Service 售后服务

Part One: Warm-up Activities

In Part One, you will practice typing sentences, paragraphs and passages dealing with after-sales service. Firstly you are supposed to read aloud and to identify the new words listed in Section A, and then try to practice typing the sentences, paragraphs and passages in Section B, C, D respectively under the teacher's guidance. While you are typing, please mark out the time you spend on each section and compare your results with your classmates.

Section A Vocabulary work

potential benefits		潜在利益
enhance	*v.*	提高;增强
honour	*v.*	使感到荣幸
definitely	*ad.*	无疑地;确实地
compatible	*a.*	协调的;一致的
host computer		(计)主机
Imp. & Exp. Corporation		进出口公司
metal embossing machine		金属浮雕机
roller	*n.*	滚筒
defective	*a.*	有缺陷的
representative	*n.*	代表
authorize	*v.*	委任;批准
overlook	*v.*	忽视;没注意到
deterioration	*n.*	变质
amicable	*a.*	友好的
dedicated	*a.*	热忱的

Section B Sentence practice

1. After-sales services of our company are satisfactory.
2. Do you provide after-sales services?
3. We will provide the customers with world-wide services.
4. After-sales services offer many potential benefits.
5. Our after-sales services for our valued customers are free.
6. After-sales services will enhance services to consumers.
7. You will be able to receive services as often as you require.
8. Please contact us by phone, fax or e-mail after you purchase our products.
9. The service centre has basic products of all supplied systems available.
10. Telephone assistance with problems and technical support is available 5 days a week, 8 hours a day.

Section C Paragraph practice

1. Our aim is to provide any kind of assistance, guidance and support to all of you that honored us by choosing our products. If you have any requires please connect to the call centre service which is available 7 days a week, 24 hours a day.
2. I bought an ACER computer last year. After using it for several months, the host computer started to make strange noises. It sounded like it was hitting other parts of the machine. How can I do?
3. China National Machinery Imp. & Exp. Corporation has ordered ten Metal Embossing Machines from an English company. Upon delivery, some rollers are found to be defective. Mr. Addison, representative of the English firm, is authorized to take up the matter with the Machinery Corporation in Beijing. He comes to discuss with Mr. Zhang from the corporation in charge of the claim.

Section D Passage practice

➤ Passage 1

After-sales services are so important that it can never be overlooked in our exportation. It also plays a major role in the organization as this department offers all buyers a service of assistance and guidance regarding their products.

In case the buyers claim for deterioration of quality, we should do our best to help them out of trouble. We should always take an earnest attitude towards it and settle the claim in an amicable way. We are dedicated to serving the buyers.

➢ Passage 2

I bought my air conditioner last year. After using it for a period of time, the machine started to make lots of noises. I called many times for service and you will never believe how they treated my call.

I told them the machine was making a lot of noises whenever I turned on my air conditioner and while I was using it. The operator told me to put the phone next to the machine and let him hear it. The reply from the operator was "…Sorry sir, but I couldn't hear a thing".

I tried to explain my situation but what he told me was they can't help me, and they can't send an engineer out because they do not know what is happening to my machine.

Disappointed, I hung up and called them again the next day. With my experience with the operator, I finally got to tell them that my machine is down and I could not turn it on. With this excuse, they finally accepted my order and said they would send an engineer out to check my machine.

Part Two: Audio-typing

Now you are going to listen to the recording. Do not refer to your textbook while you are listening. Then you are supposed to listen to each section sentence by sentence once again and type what you hear at the same time.

Section A

clip 4-25

Listen to the audio clip 4-25 and type the following short and simple sentences.

1. The quality is too bad.
2. May I speak to the sales manager?
3. Can you have it repaired?
4. I found it can't work now.
5. Could you send an engineer to have a look?
6. It is the matter with the motor.
7. Is there anything I can do for you?
8. The air-conditioner doesn't work properly.
9. Here is what differs from your sample.
10. What's the problem with the latest lot exactly?
11. A quarter of peanuts went moldy.
12. Please find out the cause.
13. We will make an investigation.

14. Let me have a look over the test report.
15. I will contact the factory at once.

Section B

Listen to the audio clip 4-26 and type the following longer and more difficult sentences.

clip 4-26

1. I'm very sorry and we are responsible for the mistake.
2. We'll exchange all merchandise that differs from our sample.
3. We guarantee the quality of the product we sold in our store.
4. We will pride ourselves on our after-sales service.
5. Once you have bought a machine you can start to use the service.
6. What we are proud of is our free after-sales service for our valued customers.
7. The factory party agreed to compensate for your damage.
8. But according to the checker's report, the packing is half-baked.
9. This is a service you will be able to have after you have finished purchasing machines with us.
10. Our Customer Service Department stays open for 7 days a week.
11. Two boxes of your product didn't coincide with your sample last time.
12. Our peanuts have enjoyed a good reputation for their superior quality for years.
13. To our astonishment, over half of peanuts are in and out the worm-eaten.
14. Yet your latest products, to our regrets, are of such poor quality that we find we must file a claim on you.
15. The quality of the second batch of goods does not meet the requirements in our contract.
16. Now show me your receipt. We cannot do a replacement without a receipt.
17. If it is because of the unqualified products, we will properly deal with it.
18. We are completely responsible for this accident. We promise we won't make this kind of mistake again.
19. I want to have your opinion on the quality problem of the goods.
20. We regret for this but we guarantee there is nothing wrong with our products.
21. We found in the examination that 20% of the goods had rusted. Please exchange them.
22. Our investigation results tell us that the factory party is responsible for the accident.
23. We customers do have the right to use after-sales services which are provided by your company.
24. Our staff in the Customer Service Department will be happy to answer all your questions and even make any amendments you require free of charge.

25. Please don't hesitate to contact us by phone, fax or e-mail for technical support after you purchase our products.

Section C

clip 4-27

Listen to the audio clip 4-27 and type the following paragraphs sentence by sentence.

1. This is a washing machine I bought in your store a month ago. You guaranteed the quality of this type of the machine at that time. But there must be something wrong with the machine. Having used it for a month, I found it can't work normally now. Could you send an engineer to have a look?
2. From the moment the sales agreement has been reached, the buyer is directed to the After-sales Department where one can receive free of charge advice and services.
3. We have received your letter of August 30, and must straightway apologize for sending you goods of inferior quality. To put the matter right we have shipped replacements for all the items you have found unsatisfactory.

Section D

clip 4-28

Listen to the audio clip 4-28 and type the following letter sentence by sentence.

➢ **A**

Dear Sirs,

We are very sorry to inform you that your last shipment is not up to your usual standard. The goods seem to be too roughly made. By another mail we have sent you a sample of this article, so that you can compare it with your original sample and see the inferiority of the goods dispatched.

We are very disappointed in this case because we have to supply these articles to new customers. We ask you to let us know immediately what you can do to help us in overcoming this difficulty.

Yours faithfully,

clip 4-29

Listen to the audio clip 4-29 and type the following letter sentence by sentence.

➢ **B**

Dear Sirs,

From your letter of April 15, you are dissatisfied with our goods because the goods were

not equal to the original quality. We write at once to say we take the responsibility for the loss and inconvenience, and it gives us the opportunity to clear the matter up immediately.

We are sending you samples of pieces we have in stock. Please select two which you consider satisfactory and return the original two pieces to us. We hope that this will settle the matter to your full satisfaction and that our friendly connection will continue as before.

Yours sincerely,

Listen to the audio clip 4-30 and type the following letter sentence by sentence.

clip 4-30

 **C**

Dear Mr. Zhang,

I have to inform you and your corporation. It's a most unpleasant incident. I must say the case is too serious to be overlooked. It is about the 500 cases of apple jam. Many tins, nearly half, have been found to lose a good deal of their quality by wetting from seawater. Our health authorities said that they were no longer suitable for human consumption. We cannot possibly deliver the goods in this condition to our customers.

You see, Mr. Zhang, the market and customers are very important for our corporation. If the matter is further delayed, we might lose our goodwill and future business. I think you won't see our position weakened, will you?

I'll cable back immediately and ask my people to send you the survey report. We hope that you replace them as soon as possible.

Yours,

Mr. Bragg

Listen to the audio clip 4-31 and type the following letter sentence by sentence.

clip 4-31

 **D**

Dear Editor,

As I was reading your article of the paper, I realized that there are some people who care what the consumers have to say. I felt like letting people know what is happening with the company's after-sales service in which I bought my computer. I think the company's after-sales service falls short of expectations. The sales and after-sales services are definitely not compatible.

Several months ago, I bought a laptop computer from the company. After using it for a few weeks, there was something wrong with its host, and the fan started to make terrible noises. I called the After-sales Department of the company again and again, and asked to send an engineer out to check my laptop computer. But they always told me that they can't help me,

they can't send an engineer out because they do not know what is happening to my machine. I finally told them that my machine is down and I cannot turn it on. With this excuse, they finally accepted my order and said they would send an engineer out to check my computer.

One morning, an engineer was at my place. After checking my computer, the engineer told me my computer definitely needed to be replaced. He told me he did not have the right spare parts to repair my machine and needed to do it another day. The engineer said that a lot of customers who bought this type of computer do give false reports in order to get the After-sales Department to send someone out to check on the computers.

With such an experience, I will make sure my family and friends do not buy anything from this company.

I really want to ask the company a question:

We customers do have the right to use the after-sales service provided that the warranty has not expired. If the company agrees to it, why ignore us?

Best regards,

A consumer

Notes

moldy	*a.*	发霉的
merchandise	*n.*	商品;货物
guarantee	*v.*	保证;担保
compensate for		赔偿
half-baked	*a.*	不周全的;不完善的
reputation	*n.*	名誉;名声
superior quality		高质量
worm-eaten	*a.*	虫蛀的;多蛀孔的
a batch of		一批
rust	*v.*	(使)生锈
free of charge		免费
straightway	*ad.*	立即;马上
inferior quality		低质量;次质量
health authorities		卫生部门
consumption	*n.*	消费
goodwill	*n.*	信誉
warranty	*n.*	保单
ignore	*v.*	忽视;不理睬

Chapter 5 Media and Publicity

媒体与宣传

5.1 Ceremonial Speeches 礼仪祝词

Part One: Warm-up Activities

In Part One, you will practice typing sentences, paragraphs and passages dealing with ceremonial speeches. Firstly you are supposed to read aloud and to identify the new words listed in Section A, and then try to practice typing the sentences, paragraphs and passages in Section B, C, D respectively under the teacher's guidance. While you are typing, please mark out the time you spend on each section and compare your results with your classmates.

Section A Vocabulary work

Myanmar		缅甸
Independence Day		独立节
millennia-old	*a.*	千年的
Auckland Lantern Festival		奥克兰元宵灯节
the 7th CISM Military World Games		第七届军运会
time-honored	*a.*	悠久的
keen perspectives and insights		真知灼见
mandate	*n.*	命令;指令
mainstream media		主流媒体
auspicious	*a.*	吉祥如意的
heartfelt thanks		衷心感谢

Auckland Council		奥克兰议会
strenuous efforts		不懈的努力
cauldron	*n.*	圣火盆
ignite	*v.*	点燃
extend warmest welcome to		表示热烈欢迎
International Olympic Committee		国际奥委会
International Military Sports Council		国际军体理事会
International Sports Federations		国际单项体育联合会
exuberance and vitality		生机活力
distinguished guest		嘉宾
International Horticultural Exhibition		世界园艺博览会
in my own name		以我个人的名义

Section B Sentence practice

1. The history of humanity is a record of one significant moment after another.
2. I have also received many sincere wishes sent by you, our friends around the world.
3. Myanmar has just celebrated its Independence Day and China will soon have its Spring Festival.
4. I am sure more and more people will come to your country and be amazed by the beauty of this millennia-old civilization.
5. It's my great pleasure and honor to participate in the Auckland Lantern Festival together with so many friends.
6. By saying so, I wish this festival a great success, and every one of you a prosperous New Year.
7. We would like to chase the Olympic dream of "Faster, higher and stronger" together with our friends.
8. I wish the 7th CISM Military World Games a full success!
9. I sincerely invite the friends from all over the world to experience China's long history and its time-honored culture.
10. I look forward to your keen perspectives and insights.

Section C Paragraph practice

1. Bearing in mind President Xi Jinping's mandate to innovate and to make good use of the opportunities provided by new media, we are speeding up our efforts to become a new kind of world-class mainstream media organization, one that is driven by innovation.

2. It warms my heart to visit Myanmar at such an auspicious time and join all of you in this delightful gathering tonight. On behalf of the Chinese government and people, let me first extend our best regards and new year greetings to people from across sectors in Myanmar.
3. I would like to take this opportunity to express my heartfelt thanks to the government of New Zealand and Auckland Council for your longtime support to the Festival and for your strenuous efforts in making the annual Festival possible and so successful.

Section D Passage practice

➢ Passage 1

Tonight, the cauldron of the 7th CISM Military World Games, a symbol of peace, development and friendship is to be ignited very soon in this stadium. At this exciting moment, on behalf of the 7th CISM Military World Games Organization Committee, I would like to extend my warmest welcome to all athletes and officials from 109 countries, and to all the distinguished guests from all over the world. I would also like to express my most sincere gratitude to the International Olympic Committee, International Military Sports Council, International Sports Federations and friends from all walks of life for your support to the 7th CISM World Games staged in Wuhan.

➢ Passage 2

By the beautiful Guishui River and at the foot of the majestic Great Wall, we are very glad to welcome all the distinguished guests to the opening of the International Horticultural Exhibition. On behalf of the Chinese government and people and in my own name, I wish to extend a warm welcome to all the guests coming to the Expo and sincere appreciation to all the friends for their support and participation.

Part Two: Audio-typing

Now you are going to listen to the recording. Do not refer to your textbook while you are listening. Then you are supposed to listen to each section sentence by sentence once again and type what you hear at the same time.

Section A

Listen to the audio clip 5-1 and type the following short and simple sentences.

clip 5-1

1. It is my great pleasure to attend the Eighth World Peace Forum.

2. It is my privilege to share this week with you.
3. I am profoundly honored to be your guests for this historic state visit.
4. It is such an honor to have you with us today.
5. First of all, let me pay tribute to our Australian friends.
6. My warm welcome goes to all the guests attending tonight's closing ceremony.
7. We will create an even better tomorrow for civilizations in Asia and beyond!
8. Going forward, we need to see where the world is going.
9. To that end, I believe it is imperative that we act in the following ways.
10. We celebrated the 70th anniversary of the founding of the People's Republic of China.
11. The Expo has provided up-close experiences with Mother Nature.
12. Have a good evening and a very good week.
13. The meeting is now drawing to a successful close.
14. I now declare the closing of the International Horticultural Exhibition 2019.
15. To conclude, I wish this conference every success!

Section B

clip 5-2

Listen to the audio clip 5-2 and type the following longer and more difficult sentences.

1. On behalf of all Americans, I offer a toast to the eternal friendship of our people and the vitality of our nations.
2. I hope you will provide each other much food for thought, build consensus, and contribute your wisdom to world peace and development.
3. Thank you for your warm welcome, your gracious hospitality, and Your Majesty's nearly seven decades of treasured friendship with the United States of America.
4. Thank you for all you do to support multilateralism and common solutions.
5. Let's keep working together for peace, prosperity and human rights for all on a healthy planet.
6. Ladies and gentlemen, I invite you all to rise and drink a toast to Mr. President and to the continued friendship between our two nations.
7. As we look to the future, I am confident that our common values and shared interests will continue to unite us.
8. We are very happy to take this opportunity to celebrate the Presidency of the Security Council.
9. Finally, I wish you a lovely evening and a very prosperous happy Chinese New Year!
10. I'm delighted to come to beautiful Davos. Though just a small town in the Alps, Davos is an important window for taking the pulse of the global economy.

11. We should join hands and rise to the challenge. Let us boost confidence, take actions and march arm-in-arm toward a bright future.
12. Let our cooperation deliver more benefits to the peoples of our five countries. Let the benefits of global peace and development reach all the people in the world.
13. It is my great pleasure to have all of you with us in the beautiful city of Xiamen, renowned as the "Egret Island".
14. On behalf of the Chinese government and people, and also in my own name, I warmly welcome all of you to the Business Forum.
15. If we take the first courageous step towards each other, we can embark on a path leading to friendship, shared development, peace, harmony and a better future.
16. I'm pleased to send you my best wishes on this auspicious occasion. In this new year, let us join hands to build a peaceful and prosperous world for all.
17. In that spirit, I wish you and your families health, success and happiness in this new year.
18. I look forward very much to the discussions and insights this week. Thank you again for this great honor.
19. This leads to the question: what shall we do to advance the new model of major country relationship between China and the US from a new starting point?
20. The answer, in my view, is to stick to the right direction of such a new model of relationship and make gradual yet solid progress.

Section C

Listen to the audio clip 5-3 and type the following paragraphs sentence by sentence.

clip 5-3

1. In this lovely season of early summer when every living thing is full of energy, I wish to welcome all of you, distinguished guests representing over 100 countries, to attend this important forum on the Belt and Road Initiative held in Beijing. This is indeed a gathering of great minds.
2. As I walked around the park with other leaders earlier in the evening, I was deeply impressed by the picturesque sceneries. I cannot help but hope that such a charming sight can be seen in more places across China and the world. It behooves us all in the international community to build and leave to our children and grandchildren a beautiful planet to live in.
3. The CIIE is an event hosted by China with the support of the WTO and other international organizations as well as a large number of participating countries. It is not China's solo show, but rather a chorus involving countries from around the world.

Section D

clip 5-4

Listen to the audio clip 5-4 and type the following passage sentence by sentence.

A

Speech by President Xi Jinping at the Asian Culture Carnival (Excerpt)

15 May, 2019

Your esteemed guests,
Ladies and gentlemen,
Dear friends,

Good evening.

In the first month of summer, the starry sky at night is magnificent. In this booming season, the guests from Asian countries together with artists and youngsters are gathering here. In the form of the Asian Culture Carnival, we celebrate this joyful cultural festival.

First of all, on behalf of the Chinese government and the Chinese people, I would love to extend our warm welcome to the guests and artists from all nations.

Flowers fill this spring garden and everything comes back to life. All Asian nations are home to ancient and splendid civilizations. They are unique in themselves and remarkable as a whole. They reinforce each other in harmony. The diversity of Asian civilizations renders the Asian cultures richer, more colorful, and more vibrant.

Tonight, such a beautiful flower of Asian culture will burst into bloom to the full. Art will transcend the ethnic boundary, connect hearts and bridge ideas, presenting to the whole world a shining Asia, a dynamic Asia and a peaceful Asia in progress.

clip 5-5

Listen to the audio clip 5-5 and type the following passage sentence by sentence.

B

Speech at the Welcoming Banquet of the Shanghai Cooperation Organization Summit in Qingdao (Excerpt)

9 June, 2018

Distinguished colleagues,
Dear guests,
Ladies and gentlemen,
Friends,

Good evening.

It is such a pleasure to have you with us here in Qingdao, Shandong Province, on the shore of the Yellow Sea. I wish to extend, on behalf of the Chinese government and people, and in my own name, a very warm welcome to all of you, particularly the state leaders and heads of international organizations who have come to attend the meeting of the Council of Heads of Member States of the SCO.

Qingdao is a famous international sailing capital. It is from here that many ships set sail in pursuit of dreams. Tomorrow, we will hold the first summit of the SCO after its expansion and draw up a blueprint for its future growth.

The Qingdao summit is a new departure point for us. Together, let us hoist the sail of the Shanghai Spirit, break waves and embark on a new voyage for our organization.

Now, please join me in a toast,

To the full success of the Qingdao Summit,

To the prosperity of our countries and happiness of our people,

To a brighter future of the SCO,

And to the health of all of you and your families,

Cheers!

Listen to the audio clip 5-6 and type the following passage sentence by sentence.

clip 5-6

 **C**

Keynote Speech at the Opening Ceremony of the First China International Import Expo (Excerpt)

5 November, 2018

Distinguished guests,

Ladies and gentlemen,

Friends,

In May 2017, I announced China's decision to hold the China International Import Expo (CIIE) starting from 2018. Today, after more than one year of preparations and with the strong support from various parties, the first CIIE is officially opened.

At the outset, I wish to express, on behalf of the Chinese government and people and also in my own name, warm welcome, sincere greetings and best wishes to you all.

The CIIE is the world's first import expo held at the national level, an innovation in the history of global trade. It is an important decision made by China to pursue a new round of high-level opening-up, and is China's major initiative to still widen market access to the rest of the world. It demonstrates China's consistent position of supporting the multilateral trading system

and promoting free trade. It is a concrete action by China to advance an open world economy and support economic globalization.

I wish all friends participating in this Expo a most pleasant and rewarding experience.

clip 5-7

Listen to the audio clip 5-7 and type the following passage sentence by sentence.

➢ **D**

Speech by the Duke of Cambridge at the International Business Festival Forum (Excerpt)

19 June, 2018

I'm personally delighted to be here in Liverpool today at the International Business Festival, and I am proud to be the event's patron for this year.

I was impressed to learn that the Festival has delivered more than half a billion pounds of new trade and investment to date. From all that I've seen, the team here have continued to build on this success, and I'd like to thank them for all their hard work.

The Business Festival is already demonstrating its success in bringing together international businesspeople to build networks and to share knowledge. There are representatives here today of businesses from across the UK and around the world—from India, from China and throughout Europe. Although our backgrounds may be different, we are all united by our shared connection through trade.

Britain has always been a champion of trade and a hub for commerce and exchange between our nations. There is perhaps nowhere in the country that embodies this more than Liverpool.

I wish you all an engaging and profitable time at the Business Festival. I hope you leave with your address books bursting with new contacts and opportunities from all across the world. I look forward to hearing about your future success.

Thank you very much.

clip 5-8

Listen to the audio clip 5-8 and type the following passage sentence by sentence.

➢ **E**

Charming Chengdu
Message by the Ambassador at the Opening Reception of "Chengdu 72 Hours Experience in London"

Distinguished guests,

Ladies and gentlemen,

Greetings to you all on behalf of the Chinese Embassy in London! I send my very warmest congratulations to the Chengdu 72 Hours Experience!

Chengdu is a city rich in history in southwestern China. It has long had a reputation as a "paradise on earth". This is a city of harmony between nature and culture. It is a city of great attractions and can boast about a unique blend of traditions and dynamism.

The charm of Chengdu lies in its natural splendor. The city has been richly endowed by nature. This means Chengdu offers a wealth of tourist destinations. These include national-rated scenic areas, natural features, with many forests and geological parks. The splendor cannot fail to win over the hearts and minds of any visitor.

Chengdu also excels in its cultural riches. With a history spanning over 2,000 years, Chengdu is among China's earliest "famous historical and cultural cities". As a result, Chengdu is the permanent seat of the "China International Festival of Intangible Cultural Heritage". Through this you will find that Chengdu is a city rich in traditions, cultural heritage and world-renowned handicrafts. For example, Chengdu is famous for its silverware, lacquerware, embroidery, bamboo and brocade crafts. In turn Chengdu offers its signature local cuisine, tea ceremonies, Sichuan Opera, shadow puppet play and, last but not least, the giant pandas.

Alongside these cultural attractions Chengdu has evolved as a key business centre. Chengdu is located right at the heart of the much desired market in China's Southwest. It has attracted over half of the Fortune 500 companies. Chengdu has now won a place on the world stage and so is making steady progress towards becoming an international metropolis.

The charm of Chengdu can go on and on! But, Chengdu 72 provides a valuable opportunity for British people to experience what the city offers. Of course, we look forward to more and more British people visiting Chengdu! That is the best way to learn about many charms of Chengdu!

Finally, I wish Chengdu 72 a great success! May the exchanges and cooperation in many fields between Chengdu and the UK yield very fruitful outcomes.

Thank you!

Notes

World Peace Forum		世界和平论坛
state visit		国事访问
pay tribute to		表示敬意
up-close	*a.*	近距离的
eternal	*a.*	永恒的

build consensus		凝聚共识
Your Majesty		陛下
multilateralism	*n.*	多边主义
Davos		达沃斯
Alps		阿尔卑斯山
take the pulse		把脉;检查
Egret Island		鹭岛
the Belt and Road Initiative		"一带一路"
solo show		个人表演
chorus	*n.*	大合唱
global economic governance system		全球经济治理体系
innovative and inclusive		创新包容的
symphony orchestra		交响乐团
starry sky		星空
the Asian Culture Carnival		亚洲文化嘉年华
be home to		某地拥有……
remarkable	*a.*	卓越的
burst into bloom to the full		尽情绽放
transcend the ethnic boundary		跨越民族界限
bridge ideas		沟通思想
the Council of Heads of Member States of the SCO		上海合作组织成员国元首理事会
dialogue partners		对话伙伴
departure point		起点
hoist the sail		扬帆起航
China International Import Expo (CIIE)		中国国际进口博览会
at the outset		首先
at the national level		国家级
major initiative		重大举措
widen market access		开放市场
multilateral trading system		多边贸易体制
free trade		自由贸易
rewarding	*a.*	有收获的
International Business Festival		国际商业节
patron	*n.*	赞助人
to date		截至今天

champion	*n.*	支持者
address book		通讯簿
opening reception		开幕式
Chengdu 72 Hours Experience in London		伦敦成都周
dynamism	*n.*	生机；活力
tourist destination		旅游景点
national-rated	*a.*	国家级的
China International Festival of Intangible Cultural Heritage		中国国际非物质文化遗产节
silverware	*n.*	银器
lacquerware	*n.*	漆器
embroidery	*n.*	刺绣(品)
bamboo and brocade crafts		竹、锦工艺品
local cuisine		地方美食
tea ceremony		茶艺
Sichuan Opera		川剧
shadow puppet play		皮影戏
international metropolis		国际化大都市

5.2 Press Conference 新闻发布会

Part One: Warm-up Activities

In Part One, you will practice typing sentences, paragraphs and passages related to press conference. Firstly you are supposed to read aloud and to identify the new words listed in Section A, and then try to practice typing the sentences, paragraphs and passages in Section B, C, D respectively under the teacher's guidance. While you are typing, please mark out the time you spend on each section and compare your results with your classmates.

Section A Vocabulary work

open the floor for questions		回答提问
stimulus	*n.*	刺激措施
market-oriented reform		市场化改革
market entity		市场主体
trend-bucking surge		逆势增长
online shopping		网购

express delivery service		快递
teleworking		云办公
be attributable to		归因于
supply-side structural reform		供给侧结构性改革
new drivers of growth		新动能
start-up	*n.*	创业公司
Asahi Shimbun		(日本)《朝日新闻》
take a big hit on		严重打击
FTA (Free Trade Agreement)		自由贸易协定
inter-flow	*n.*	交流
resumption of economic activities		恢复经济活动;复工复产
geographical proximity		毗邻
as scheduled		如期
multipronged	*a.*	多种的;多管齐下的
subsistence allowance		低保
unemployment benefit		失业保障

Section B Sentence practice

1. Friends from the media, good afternoon.
2. With these words, I'm happy to open the floor for questions.
3. Can we expect China to deliver more ambitious stimulus in the months to come?
4. This is in keeping with market-oriented reform.
5. It is essential that we keep China's economic growth on a steady course.
6. China will remain a positive force driving global economic recovery and growth.
7. How does China view the post-COVID-19 world and the future of globalization?
8. Not setting a specific GDP growth target does not mean that economic development is not important.
9. Such a target was set on the basis of the situation on the ground.
10. The delivery of these measures will have to be recognized by businesses and our people.

Section C Paragraph practice

1. We must work hard to help more new market entities emerge. We have seen a trend-bucking surge of growth in new forms of business and industry during the COVID-19 response, including online shopping, express delivery services, and teleworking. Some new forms of business have even seen their revenues increase by two thirds.

2. The world will certainly not be the same again; history always moves forward. Throughout world history, humanity has progressed by wrestling with one disaster after another. In China's view, if countries make the right choice and stay on the right path, the world will triumph over the virus and embrace a brighter future.
3. I believe such developments are very much attributable to our initiatives and reforms in recent years, including the supply-side structural reform, our efforts to promote high-quality development and foster new drivers of growth, and our initiative of encouraging business start-ups and innovation. We must use the experience gained in this process to the fullest extent to foster more growth drivers and help more market entities grow.

Section D Passage practice

➢ Passage 1

Asahi Shimbun: COVID-19 has taken a big hit on the global economy. But China has been successful in bringing the spread of the virus under control. What does China plan to do to advance economic cooperation with Japan and other neighboring countries?
Premier Li: I recall that last year at the leaders' meeting of East Asian cooperation, leaders of 15 countries made the commitment of signing the RCEP by the end of this year. I hope and believe that this commitment will not come to nothing. China, Japan and the ROK are also working closely together to advance their FTA development. The three countries are close neighbors, and we would like to work with the other two countries to develop a mini economic cycle within the bigger economic cycle. For example, recently China and the ROK have opened a fast-track service for personnel inter-flows in areas such as commerce and technology. We believe this will benefit the resumption of economic activities, and our geographical proximity has put us in a good position to benefit from this earlier.

➢ Passage 2

People's Daily: This year, China plans to win its battle against poverty. But because of COVID-19, many families have seen a decline in their household income. And some are even at the risk of falling back into poverty. So are we able to fulfill the task of winning the battle against poverty this year? And what will the government do to meet people's essential needs?
Premier Li: This year, we are determined to end poverty as scheduled. Before COVID-19 struck, there were some 5 million people living below the poverty line. But because of the disease, some may have fallen back into poverty. Hence, we now face a greater task in meeting our goal. But with our multipronged policies and measures to ensure the essential needs of our people, we have the confidence to win the battle against poverty this year.

Governments at all levels must always put people's interests first and bear in mind the hardships of the Chinese people. In introducing each and every policy, we must make sure that it contributes to the well-being of all families and to the better lives of our people. In this respect, we have decided to expand the coverage of subsistence allowance and unemployment benefits. We must fully deliver all our commitments.

Part Two: Audio-typing

Now you are going to listen to the recording. Do not refer to your textbook while you are listening. Then you are supposed to listen to each section sentence by sentence once again and type what you hear at the same time.

Section A

clip 5-9

Listen to the audio clip 5-9 and type the following short and simple sentences.

1. This year's press conference is held at a special time.
2. Since we have limited time, I invite you to be direct with your questions.
3. We must take strong measures to cope with the current downward economic pressure.
4. We must keep our policies stable and ensure their continuity.
5. We will generate tremendous creativity in this process.
6. To address this problem, we will make further tax deductions.
7. Our end goal is to deliver concrete benefits to companies and market entities.
8. There are still complaints about some issues concerning quality of life.
9. What specific measures will be adopted to improve China's business environment?
10. Serving the real economy is the bounden duty of the financial sector.
11. What are your measures to resolve these issues?
12. In the meantime, we also need to forestall financial risks.
13. The new coronavirus outbreak has devastated economies around the world.
14. What measures should be adopted to better regulate the growth of sharing economy?
15. China not only promoted the global economic recovery, but also contributed to world peace.

Section B

clip 5-10

Listen to the audio clip 5-10 and type the following longer and more difficult sentences.

1. Our decision is designed for economic growth to deliver more real gains to our people

and promote higher quality development in China.

2. China will not waver in this commitment, nor is it possible for us to shut our door to the outside world.
3. With a particular focus on supporting jobs and people's livelihoods, our people will have money to spend and consumption will drive market vitality.
4. Money invested in the people will generate new wealth, help us protect and preserve the tax base and make public finance more sustainable.
5. We raised the deficit ratio for this year by 0.2 percentage point to 2.8 percent, which is below the international warning line of 3 percent.
6. Facing new circumstances, we will stay firmly grounded in China's realities and take a long-term view.
7. For other industries, we will also work to ensure that the tax burden on companies will only go down, not up.
8. We also need to ensure that government spending in key areas related to people's lives and in fighting the three critical battles will increase.
9. Through 40 years of reform and opening-up, China has made remarkable achievements, delivering benefits to its entire population.
10. We will continue to develop our socialist market economy, and pursue market-oriented reforms.
11. Our goal is to further cut the financing cost for small and micro companies by another one percentage point this year.
12. Strengthening financial services and preventing financial risks are mutually reinforcing.
13. E-commerce and express delivery services have made it possible for industrial goods to reach rural areas.
14. Investment from Hong Kong and Macao accounts for 70 percent of all overseas investment on the mainland.
15. Although China currently has a surplus, more than 90 percent of Chinese companies' profits were taken by the United States.
16. Do you think that the Chinese economy can still remain the global economic engine driving the global economy while global economic growth is sluggish?
17. If we are able to achieve the 6.5 percent target this year, it will generate more economic output than last year.
18. Our fiscal deficit is less than 3 percent and the capital adequacy ratio of commercial banks stands at 13 percent.
19. Like many other countries, China is the beneficiary of globalization as China is consistently advancing its opening-up.

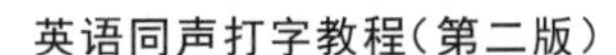

20. We proposed to establish a free trade zone or make negotiations on investment and trade treaties with many other countries.

Section C

clip 5-11

Listen to the audio clip 5-11 and type the following paragraphs sentence by sentence.

1. Last year China took a number of measures to ease monetary conditions. China also cut taxes and fees. This year China is promising more monetary easing, more tax cuts and more infrastructure spending.
2. Globalization represents an inevitable trend in the development of the world and a strong tide driving human progress. It has turned the global economy into an ocean, to which every river flows.
3. Just now, I talked more about boosting consumption. That doesn't mean investment is not important. We will also expand effective investment. There will be an increase of 1.6 trillion RMB yuan of special local government bonds this year, and some treasury bonds as well.

Section D

clip 5-12

Listen to the audio clip 5-12 and type the following passage sentence by sentence.

➢ A

Phoenix TV of Hong Kong: Mr. Premier, in your government work report this year, you said that efforts will be made to develop the Guangdong-Hong Kong-Macao Greater Bay Area. But meanwhile, some people in Hong Kong worry about whether this will cause Hong Kong to lose its unique features, or whether it will affect the implementation of "One Country, Two Systems," even blur the line between the two systems. How do you respond to that?

Premier Li: We want to build the Guangdong-Hong Kong-Macao Greater Bay Area into a world-class city cluster, with these three places complementing each other's unique comparative strengths. Otherwise, this Greater Bay Area would not be competitive internationally. The outline for the development program has been formulated and will soon be made public for implementation.

We encourage Hong Kong and Macao to integrate their own development into the overall national development. In this process, we will uphold the principle of "One Country, Two Systems," under which the people of Hong Kong administer Hong Kong, and the people of Macao administer Macao, with both regions enjoying a high degree of autonomy. We have

confidence that they will draw upon each other’s comparative strengths，and work together in building a strong new pole of growth.

Thank you.

Listen to the audio clip 5-13 and type the following passage sentence by sentence.

clip 5-13

 **B**

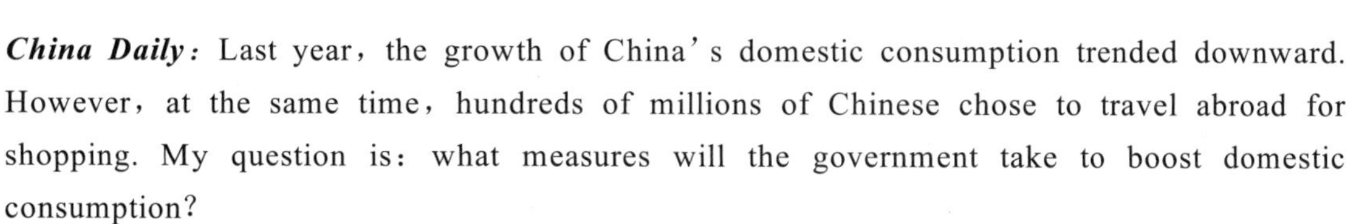

***China Daily*：** Last year，the growth of China’s domestic consumption trended downward. However，at the same time，hundreds of millions of Chinese chose to travel abroad for shopping. My question is：what measures will the government take to boost domestic consumption?

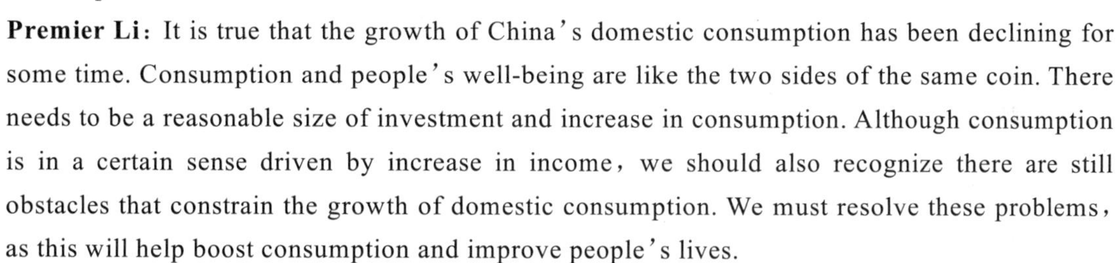

Premier Li： It is true that the growth of China’s domestic consumption has been declining for some time. Consumption and people’s well-being are like the two sides of the same coin. There needs to be a reasonable size of investment and increase in consumption. Although consumption is in a certain sense driven by increase in income，we should also recognize there are still obstacles that constrain the growth of domestic consumption. We must resolve these problems，as this will help boost consumption and improve people’s lives.

Listen to the audio clip 5-14 and type the following passage sentence by sentence.

clip 5-14

 **C**

CCTV： In the government work report，this year’s economic policies have been spelled out. What will the government do to ensure that all these funds will be truly delivered to benefit companies，instead of just circulating within the financial sector，and bring real gains to the general public?

Premier Li： As I said just now，our policies，which are of a sizable scale，are designed to provide vital relief to businesses and revitalize the market，with particular focus on stabilizing employment and ensuring people’s livelihood. They are not focused on large infrastructure projects. This is because big changes have taken place in China’s economic structure，where consumption is now the primary engine driving growth，and micro，small and medium-sized companies now provide over 90 percent of all jobs in China. Under the sizable-scale policies introduced this time，some 70 percent of the funds will be used to support the increase in people’s income through direct or relatively direct means in order to spur consumption and energize the market.

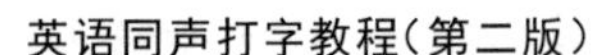

clip 5-15

Listen to the audio clip 5-15 and type the following passage sentence by sentence.

D

China Daily: In this year's government work report, the target of new urban jobs has been revised downward, and the target of surveyed urban unemployment rate upward, compared with the levels last year. In the face of the severe employment situation, what will the government do to avert massive job losses and help college graduates and rural migrant workers find jobs?

Premier Li: We have set this year's target of new urban jobs at over 9 million, somewhat below last year's level. To attain this goal, we need to maintain a certain level of economic growth. The urban surveyed unemployment rate target is set at around 6 percent. In fact, in April, that figure already hit 6 percent, so such a target was set on the basis of the situation on the ground.

Employment matters the most in people's lives. It is something of paramount importance for all families. Many self-employed individuals have seen their businesses grinding to a halt for several months. And some export companies have also been in great difficulty for lack of orders. This is affecting the jobs of their employees. We must provide support to all these people. But most importantly, we must help them land jobs. There is a labor force of 900 million in China. If they are out of work, there will be 900 million mouths to feed; if they are all put to work, 900 million pairs of hands will be able to generate tremendous wealth.

Notes

downward economic pressure		经济下行压力
continuity	*n.*	连续性
tax deduction		减税
real economy		实体经济
bounden duty		天职
forestall	*v.*	防范
sharing economy		共享经济
waver	*v.*	动摇
tax base		税源
deficit ratio		赤字率
warning line		警戒线
financing cost		融资成本
mutually reinforcing		相辅相成的

surplus	*n.*	顺差
sluggish	*a.*	疲软的
fiscal deficit		财政赤字
capital adequacy ratio		资本充足率
commercial bank		商业银行
beneficiary	*n.*	受益者
free trade zone		自由贸易区
inevitable trend		必然趋势
treasury bond		国债
Phoenix TV of Hong Kong		香港凤凰卫视
the Guangdong-Hong Kong-Macao Greater Bay Area		粤港澳大湾区
blur the line		模糊界线
high degree of autonomy		高度自治
new pole of growth		新的增长极
trend downward		下滑
provide vital relief		纾困
revitalize the market		激发市场活力
infrastructure project		基建项目
spur consumption		促进消费
avert massive job losses		遏制失业潮
rural migrant worker		农民工
self-employed individual		个体工商户
grind to a halt		逐渐停止
land jobs		就业

5.3 Advertising and Publicity 广告宣传

Part One: Warm-up Activities

In Part One, you will practice typing sentences, paragraphs and passages dealing with advertising and publicity. First you are supposed to read aloud and to identify the new words listed in Section A, and then try to practice typing the sentences, paragraphs and passages in Section B, C, D respectively under the teacher's guidance. While you are typing, please mark out the time you spend on each section and compare your results with your classmates.

Section A Vocabulary work

exhibitor	*n.*	参展商
high-speed railway		高铁
gastronomy	*n.*	美食
drainage area		流域
conference and exhibition center		会展中心
the Revolution of 1911		辛亥革命
the middle reaches of the Yangtze River		长江中游
thoroughfare to nine provinces		九省通衢
logistic and cargo distribution center		物流及货物配送中心

Section B Sentence practice

1. We really value the development of 5G technology.
2. The trade show covered culture, tourism, finance, sports, robotics and 5G.
3. Morning and night markets could generate about 100 million jobs.
4. More than 18,000 companies from 148 countries and regions joined.
5. Overseas exhibitors and guests mainly attended the fair through online platforms.
6. Indeed, traditional markets are a rich source of employment.
7. Local markets can respond to local demands rapidly.
8. Advancement in digital technology also eases traffic congestion in Hangzhou.
9. China is leading the world in the building of high-speed railways.
10. Bike sharing has seen explosive growth in China in the past few years.

Section C Paragraph practice

1. Tourism development has prospered since Macao's reunification with the motherland, and leisure and entertainment, cultural heritage and gastronomy were the most renowned tourism sectors.
2. Xi'an, the capital of Shaanxi Province, is a new industrial base and scientific and educational center in China. It is the hub of communications between eastern and western China and is an important city in Northwest China. Xi'an is an important center for the origin of ancient civilization in the drainage area of the Yellow River.
3. Hong Kong is an important conference and exhibition center in Asia Pacific Region, and always plays as the bridge for economic and trade exchanges between the mainland and overseas areas. There are more than 80 exhibitions and 420 conventions held here every year, attracting more than 20,000 exhibitors, 37,000 delegates and 3.3 million visitors from all over the world.

Section D Passage practice

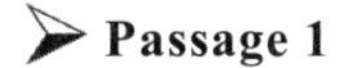

Passage 1

Welcome to Wuhan

Wuhan, capital of Hubei Province, is the largest city in central China and covers an area of 8,569 square kilometers with a population of 12 million. Wuhan is known as the "city of river". The world's third longest river—the Yangtze River and its greatest branch, the Hanshui River flow across the city and divide it into three parts, namely Wuchang, Hankou and Hanyang.

Wuhan is a famous historical and cultural city with the history of 3,500 years. It is one of the birthplaces of the brilliant ancient Chu Culture in China and the birthplace of the Revolution of 1911. Located in the middle reaches of the Yangtze River, Wuhan has been regarded as the "thoroughfare to nine provinces". It has long been an important economic, industrial and transportation hub for the country. It is also the largest inland logistic and cargo distribution center in China.

Welcome to Wuhan!

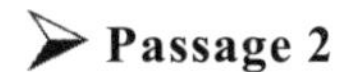

Passage 2

ABC China CEO Talks CIIE & Chinese Beauty Market

This year, we will have more brands and more innovation. For ABC, we believe that the future will be for the beauty tech. Because technology will allow beauty more personalized. And this is what we see very much this year, and you will see a lot of beauty tech within our presence.

The personalization today is the new area of development in China. In the past, we were saying there are one thousand women in one look, and today is one thousand look for one woman. The fact is that every day somebody wants to feel different, and everybody needs to feel unique. This is where beauty brands and beauty services can answer to that. In this context, healthy competition makes this market growing.

Part Two: Audio-typing

Now you are going to listen to the recording. Do not refer to your textbook while you are listening. Then you are supposed to listen to each section sentence by sentence once again and type what you hear at the same time.

Section A

clip 5-16

Listen to the audio clip 5-16 and type the following short and simple sentences.

1. The tide of reform and opening up has brought the city into the opening front.
2. Hangzhou is an important city in China's coastal areas.
3. I came to Beijing and invested in an e-commerce company.
4. Top universities will find a home here and so will our future R&D center.
5. Green and sustainable development has been a top priority.
6. Online shopping has changed the way people do shopping in China.
7. China is entering a cash-free society.
8 Xi'an has a relatively advanced industry.
9. It has provided 8.5 billion U.S. dollars in financing.
10. All countries are sharing the opportunities of economic growth.
11. WeChat's mobile payment success is also drawing the attention of Facebook.
12. Shanghai has huge potential in the development of financial services.
13. The current mobile payment tools will still be the dominant force.
14. 5G has now gradually permeated into people's everyday lives.
15. China's mineral resources are widely distributed throughout the country.

Section B

clip 5-17

Listen to the audio clip 5-17 and type the following longer and more difficult sentences.

1. By this way, business cost is reduced, the quality of services is improved, and the consumers are benefited.
2. It is also one of the country's famous production regions of tea, grain and timber and a cradle of Chinese revolution.
3. Macao has been a bridge for exchanges between Chinese and Western cultures for more than 400 years, forming a unique blend of historical and cultural heritages.
4. Various robots were on display showing skills in cooking, delivering goods and performing surgeries.
5. The fair was the first large global trade event held both online and offline by China since the COVID-19 outbreak.
6. We're going to see artificial intelligence do more and more, and as this happens costs will go down, outcomes will improve, and our lives will get better.
7. Hangzhou is a world-famous tourist destination, a city with long history and profound

culture, and also one of the seven ancient capitals in China.

8. The Macao Special Administrative Region is situated at the Pearl River Delta on the southeast coast of Mainland China.
9. As China continues to open up, more and more foreign investors are coming here to do business.
10. Long history and advanced culture have endowed the city with numerous world-famous places of historical interest and scenic beauty.
11. The Asian Infrastructure Investment Bank (AIIB) is a multilateral development bank with a mission to improve social and economic outcomes in Asia.
12. Headquartered in Beijing, we began operations in January 2016 and have now grown to 100 approved members worldwide.
13. By investing in sustainable infrastructure in Asia and beyond, we will better connect people, services and markets.
14. We have created more and more opportunities for Asian and global economies by strengthening infrastructure development.
15. Consumers in China transacted nearly 280 trillion yuan in mobile payments alone last year.
16. One popular mobile payment app in China now has 800 million monthly active users across the nation.
17. Shanghai's development as an international financial center calls for concerted efforts of all parties.
18. We will continue to speed up the construction of the 5G network, optimize the environment for innovation, and inject new energy into economic growth.
19. With rapid economic development, China needs a large number of mineral products and relevant energy and raw material products.
20. Many foreign friends have discovered four new great inventions in China, which are high-speed railway, mobile phone payment, bike sharing and online shopping.

Section C

Listen to the audio clip 5-18 and type the following paragraphs sentence by sentence.

clip 5-18

1. We are entering an age of accelerated development of artificial and robotic technology. Digital machines have escaped their narrow confines and started to demonstrate broad abilities in pattern recognition, complex communication, and other domains that used to be exclusively human.
2. Founded in 1879, the Canadian National Exhibition (CNE) is one of the largest

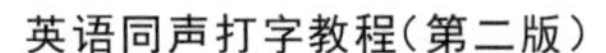

annual fairs in North America, which attracts over 1.4 million visitors during its operation. CNE is a large employer of students and young people, and has firmly established itself as Toronto's annual summer celebration.

3. The concept of digital economy is stressing that our current economy is significantly influenced by information technology. New ways of working, new means of communication, new goods and new services, and new forms of community have come into being with the development of the digital technology or information technology.

Section D

clip 5-19

Listen to the audio clip 5-19 and type the following passage sentence by sentence.

A

Shanghai Disney Resort

Shanghai Disney Resort, the first Disney resort in China's Mainland, is a place where friends and families can escape together to a whole new world of fantasy, imagination, creativity and adventure.

The resort is home to the Shanghai Disneyland theme park, featuring seven lands, as well as two themed hotels—Shanghai Disneyland Hotel and Toy Story Hotel, Disney town, a large shopping, dining and entertainment district, a Broadway-style theatre, Wishing Star Park and other outdoor recreation areas.

Shanghai Disneyland is a Magic Kingdom-style theme park featuring classic Disney storytelling and characters but with authentic cultural touches and themes tailored specifically for the people of China.

Shanghai Disney Resort offers something for everyone—thrilling adventures, lush gardens where guests can relax together, and enriching interactive experiences, all with the world-class guest service that Disney is known for around the globe.

clip 5-20

Listen to the audio clip 5-20 and type the following passage sentence by sentence.

B

Huawei Technologies CO., LTD.

Good morning, everyone!

Founded in 1987, Huawei is a leading global provider of information and communications technology (ICT) infrastructure and smart devices. We are committed to bringing digital to every person，home and organization for a fully connected，intelligent world.

We have nearly 188，000 employees，and we operate in more than 170 countries and regions，serving more than three billion people around the world. Together with our partners，we provide innovative and secure network equipment to telecom carriers. We provide our industry customers with open，flexible，and secure ICT infrastructure products. In addition，we provide customers with stable，secure，and trustworthy cloud services that evolve with their needs. With our smartphones and other smart devices，we are improving people's digital experiences in work，life，and entertainment.

We will make the most of this historic opportunity，and boldly forge ahead to build a fully connected，intelligent world.

Thank you.

Listen to the audio clip 5-21 and type the following passage sentence by sentence.

clip 5-21

 **C**

The Special Economic Zone of Shenzhen

Shenzhen used to be a small fishing village. However，it took off economically 40 years ago and was designated as the first special economic zone（SEZ）in China in 1980.

Hailed as "China's Silicon Valley"，it is also now the center of China's tech industry and home to such global companies as Huawei and Tencent. Currently，Shenzhen is building toward its future and aims to be a global model that reflects a sense of competitiveness and innovation.

Thanks to policy support，Shenzhen has created a free and fair business environment，and become a fertile ground for innovation. It has also provided the most salient lessons for other places in China.

Listen to the audio clip 5-22 and type the following passage sentence by sentence.

clip 5-22

 **D**

About Hong Kong

Hong Kong，once described as a "barren rock"，has become a world-class financial，trading and business center and，indeed，a great world city.

Hong Kong has no natural resources，except one of the finest deep-water ports in the world. A hardworking，adaptable and well-educated workforce of about 3.6 million，coupled with entrepreneurial spirit，is the foundation of Hong Kong's productivity and creativity.

Situated at the southeastern tip of China，Hong Kong is ideally positioned at the center of rapidly developing East Asia. With a total area of about 1,100 square kilometers, it mainly covers Hong Kong Island, the Kowloon peninsula just opposite, and the New Territories.

Hong Kong is the world's famous trading economy, foreign exchange market and banking center, as well as Asia's 2nd biggest stock market. Hong Kong is one of the world's top exporters of garments, watches and clocks, toys, games, electronic products and certain light industrial products.

Notes

permeate	*v.*	渗透
artificial intelligence		人工智能
Pearl River Delta		珠江三角洲
Asian Infrastructure Investment Bank (AIIB)		亚洲基础设施投资银行(亚投行)
mobile payment		移动支付
WeChat Pay		微信支付
inject new energy into		注入新活力
digital economy		数字经济
pattern recognition		模式识别
Canadian National Exhibition(CNE)		加拿大国家展览会
devalue	*v.*	贬值
Disney Resort		迪士尼度假区
Toy Story Hotel		玩具总动员酒店
Broadway-style	*a.*	百老汇风格的
Wishing Star Park		星愿公园
outdoor recreation area		户外休闲区域
lush	*a.*	郁郁葱葱的
information and communications technology (ICT)		信息通信技术
telecom carrier		电信运营商
cloud service		云服务
special economic zone (SEZ)		经济特区
fertile ground		沃土
salient	*a.*	突出的
deep-water port		深水港
entrepreneurial spirit		创业精神

southeastern tip		东南端

5.4. Business News 商务新闻

Part One: Warm-up Activities

In Part One, you will practice typing sentences, paragraphs and passages dealing with business news. Firstly you are supposed to read aloud and to identify the new words listed in Section A, and then try to practice typing the sentences, paragraphs and passages in Section B, C, D respectively under the teacher's guidance. While you are typing, please mark out the time you spend on each section and compare your results with your classmates.

Section A Vocabulary work

retail investor		散户
outstrip	*v.*	超出
default on		拖欠
fragility	*n.*	脆弱性
Anta		安踏
in bulk		批量
bargain hunter		专买便宜货者
trigger	*n.*	导火索
downturn	*n.*	衰退
the Depression		(20世纪30年代的)大萧条
walled-off	*a.*	封闭的
indices	*n.*	指数(index的复数形式)
dive	*n.*	暴跌
talent pool		人才库
scramble	*v.*	争抢
Microsoft Research		微软研究院
churn	*n.*	搅乳器;动荡
Prudential Financial		保德信金融集团
reskill	*v.*	再培训
upskill	*v.*	技能提升
slam	*v.*	猛烈抨击
bouquet	*n.*	花束
foliage	*n.*	叶子

go viral　　疯传;爆红
spark controversy　　引发争议

Section B Sentence practice

1. An official digital currency could help address a risk from this transition.
2. China appears to be in much better economic shape than other large economies.
3. Demand for shares from retail investors outstripped supply by 1,148 times.
4. As recession bites, people may default on car loans.
5. The pandemic is exposing our food system's fragility.
6. Half of all work tasks will be handled by machines by 2025.
7. A pair of Anta shoes typically costs a third less than a similar pair of Nike shoes.
8. Can the stalls really help the economy?
9. Products are cheaper if you buy in bulk with fellow bargain hunters.
10. China aims to launch the world's first official digital currency.

Section C Paragraph practice

1. Nongfu Spring is the industry leader. It accounted for 29% of the volume sold in China in 2019. Foreign brands are easily spotted in many Chinese supermarkets. But none has a market share greater than 6.5%.
2. Though the housing market has not been the trigger of economic woes this time, investors and homeowners still braced for the worst as it became clear that Covid-19 would push the world economy into its deepest downturn since the Depression of the 1930s.

Section D Passage practice

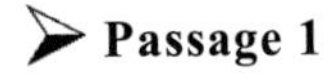

Passage 1

A Big Dive into the Talent Pool

As the pandemic eases in the United States, many employers are scrambling to find workers. In April, a record number of Americans quit their jobs. Worldwide, 40% of employees are ready to resign, according to Microsoft Research. Many seek a better work-life balance or simply more pay. But amid this workplace churn, companies are also recognizing what already exists in their current employees: an eagerness and potential to learn new skills in order to create new opportunities.

A survey in May by Prudential Financial found nearly half of American workers say the pandemic has caused them to reevaluate their skill sets. About 1 in 5 have put a greater priority on pursuing education or learning a new skill. Of those that plan to leave their jobs, 6 in 10 have sought training on their own since the start of the pandemic.

Overall, more than two-thirds of U.S. companies began last year to invest in reskilling or upskilling to deal with the effects of the pandemic, according to a survey by an online training company. One reason is that it can cost far more to recruit a new employee compared with the cost of reskilling an internal employee.

➢ **Passage 2**

Supermarket Sells Autumn Leaves

UK-based supermarket Waitrose has been slammed by social media users for selling bouquets of "autumn seasonal foliage" at £6 apiece.

Photos of various tree leaves packaged in transparent plastic and arranged as flower bouquets went viral on social media recently, sparking controversy because of the product's price, six British pounds.

Considering tree leaves can literally be picked up from the street this time of year, it's understandable that some people went after Waitrose, accusing the supermarket of trying to make money by selling something that is actually free.

Despite the generally negative feedback to its autumn-themed product, the British supermarket explained its decision to charge £6 for a bunch of tree leaves by claiming that it was inspired by a demand from shoppers.

Interestingly, there were those who seemed to somewhat justify the price tag of this bag of leaves by saying that it was obviously "high quality foliage".

Part Two: Audio-typing

Now you are going to listen to the recording. Do not refer to your textbook while you are listening. Then you are supposed to listen to each section sentence by sentence once again and type what you hear at the same time.

Section A

Listen to the audio clip 5-23 and type the following short and simple sentences.

clip 5-23

1. One reason for Nongfu's success is its effort to cater to all market segments.
2. iQiyi offers customers a deep catalogue of licensed and original content.
3. China's bustling digital economy has spawned thousands of start-ups.
4. Commercial farming in the land-scarce city was phased out in the 1970s and 1980s.

5. Urban and suburban property prices are rising at roughly the same pace.
6. Investors were deeply pessimistic about the car industry.
7. Big Western firms are starting to re-open some plants.
8. Relying on market forces has created many of the problems we now face.
9. Adidas saw sales in China drop by 58% in the same period.
10. Central to the firm's ascent is the concept of social shopping.
11. Remote working is not possible for everyone, of course.
12. Government subsidies make excellent fertiliser.
13. Delivery apps have transformed urban life in China.
14. China might find it easier to make nominal interest rates negative.
15. The biggest companies in the world base their businesses on data.

Section B

clip 5-24

Listen to the audio clip 5-24 and type the following longer and more difficult sentences.

1. Share prices of developers and property-traders fell by a quarter in the early phase of the pandemic, but have recovered much of the fall.
2. Car firms have high fixed costs, so when they run below capacity, they lose money fast.
3. A "robot revolution" would create 97 million jobs worldwide but destroy almost as many, leaving some communities at risk.
4. Rising disposable incomes and public anxiety about the safety of tap water have fueled demand among Chinese for the bottled variety.
5. Studies suggest a strong link between falling real interest rates and higher house prices.
6. This could be an opportunity to diversify our supply chain, promote sustainable agriculture and benefit local businesses.
7. The top eight retailors account for more than 90% of all grocery sales in Britain, with Tesco alone accounting for 27%.
8. Employers seem less impressed by a job applicant's formal education but more by his capability to learn new skills.
9. Over the past several decades, women have marched into male-dominated industries such as medicine, law and finance.
10. The pace of change in industries will require more than half of employees to acquire new skills by 2025.
11. Firms there started setting up street stalls in March, creating more than 100,000 jobs.
12. Some countries are imposing limits on exports of staples to ensure they can feed their

own populations.

13. The shock of unemployment may push men into jobs traditionally held by women, study shows.
14. Overall, just 12% of the workers surveyed wanted to return to a normal office schedule.
15. Routine or manual jobs in administration and data processing were seriously threatened by automation.
16. Suzhou and Shanghai, among other cities, have recently opened glitzy outdoor night markets.
17. Today just 720 square kilometres of land, less than 1% of Singapore, is set aside for farms.
18. Dozens of central banks have started looking at whether to issue digital currencies.
19. The progress the company had made allowed management to focus again on longer-term efforts to achieve carbon neutrality.
20. Online retailers use algorithms to respond instantly to fluctuations in supply and demand.

Section C

Listen to the audio clip 5-25 and type the following paragraphs sentence by sentence.

clip 5-25

1. Nissan has studied various ways to repurpose batteries so they can be used to generate renewable energy in homes and buildings as well as for emergency power during natural disasters. It has already started rolling out similar initiatives in the UK and the US. But the company is expected to strengthen its efforts as it seeks to electrify all of its new vehicles in core markets by the early 2030s.
2. China's four largest commercial banks began internal tests this month. The city of Suzhou will give some digital currency to government employees next month to cover transportation costs, according to state media.
3. In August house prices in Germany were 11% higher than the year before; rapid growth in South Korea and parts of China has prompted the authorities to tighten restrictions on buyers. Three factors explain this strength: monetary policy, fiscal policy and buyers' changing preferences.
4. Alibaba and Tencent continue to lord it over digital China. With market capitalisations of nearly $700bn apiece, they are the world's seventh- and eighth-biggest listed companies, respectively.

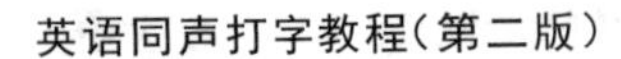

Section D

clip 5-26

Listen to the audio clip 5-26 and type the following passage sentence by sentence.

➢A

Are People Loyal to Brands?

You might have heard the term "brand loyalty". But why do people feel compelled to stick with brands? Brand loyalty is described as a positive feeling that consumers identify with a certain product or company. Huge companies pump millions into marketing and advertising, using analytics to determine who is their ideal customer profile. It's no surprise that the ads which pop up on social media somehow feel targeted at or tailored for you. This is because companies spend a lot of time and money analysing who is most likely to become loyal customers.

This extends to the supermarkets. Many believe that the more expensive branded products are much better than the supermarkets' own brand. Money-saving experts like Martin Lewis, encourage us to give up the premium or branded products and buy the value versions.

So, are we really brand loyal? Companies certainly want us to be. But there are those who believe it's more to do with brand habit—that feeling of comfort you get from buying the same product over and over again. Once we're familiar with a brand and we know that it's OK, we don't feel compelled to try anything else. So, the next time you find yourself buying your favourite brand, it might not be down to brand loyalty, but rather to habit, or even that you have been targeted by a specific company through tailored ads.

clip 5-27

Listen to the audio clip 5-27 and type the following passage sentence by sentence.

➢B

Change is Coming, Whether the Oil Industry Likes It or Not

In a dramatic double whammy this week, Royal Dutch Shell and ExxonMobil, two enormous oil companies, each lost a high-profile fight against climate activists. The consequence is that two boardrooms may be forced to overturn their business strategies in favor of greener investments over the next decade.

Change is coming, whether the industry likes it or not. The question is whether polluting businesses will engage productively in the process of decarbonization, or whether they will fight a futile rear-guard action to preserve as much of the unsustainable status quo as they can.

The world still needs oil. It is neither feasible nor desirable to end the oil business. But it is

even more important for the future of humanity that polluting industries should not engage in their own forms of reality denial.

The science of climate change is clear, severe consequences are already visible, and rising generations will not tolerate inaction. Society will force businesses to face up to climate change sooner than they may imagine, and it is in their best interest to advocate aggressively for policies that make the forthcoming energy transition predictable, economically rational and effective.

Listen to the audio clip 5-28 and type the following passage sentence by sentence.

clip 5-28

 C

Coronavirus Will Hasten the Decline of Cash

The dramatic drop in cash is the obvious result of more people having to do their shopping online. When they do venture out, many are suspicious of handling cash, which others might have touched.

Traders who used to turn up their noses at plastic are bringing out their card readers. Once people change to buying over the Internet or paying by card, they tend to carry on.

The problem is the faster the switch away from cash happens, the bigger the danger that shops stop taking it and more cash machines are closed. That could make life more difficult for the millions who still rely on notes and coins to run their lives.

Listen to the audio clip 5-29 and type the following passage sentence by sentence.

clip 5-29

 D

China's World-beating Growth Rate of 3.2%

On July 16 China reported that GDP grew by 3.2% in the second quarter compared with a year ago, rebounding from its coronavirus lockdown. This makes it, by far, the best-performing big economy.

Traffic congestion returned as people went back to work, partly because, wary of public transport, more commuted by car. Banks ramped up their lending to keep businesses afloat. Some credit flowed into the property market.

Yet flights are still down as few people go on trips. They also avoid crowds, taking the subway less often. Spending on restaurants, including takeaways, is weak, which in turn points to the soft labour market.

That is all to say: China's rebound from the coronavirus crisis is impressive, but it is not yet back to normal.

Notes

market segment		细分市场
bustling	*a.*	蓬勃发展的
spawn	*v.*	催生
land-scarce	*a.*	土地稀缺的
be phased out		被逐渐淘汰
ascent	*n.*	崛起
subsidy	*n.*	补贴
nominal interest rate		名义利率
developer	*n.*	开发商
fixed cost		固定成本
capacity	*n.*	产能
disposable income		可支配收入
fuel	*v.*	推动
Tesco		乐购(全球三大零售商之一)
scarcity	*n.*	匮乏
street stall		摊点
staple	*n.*	主食
glitzy	*a.*	盛大的
central bank		中央银行
carbon neutrality		碳中和
algorithm	*n.*	算法
fluctuation	*n.*	波动
repurpose	*v.*	改换用途
renewable energy		可再生能源
emergency power		应急电源
roll out		推出(新产品、服务等)
electrify	*v.*	使……电气化
prompt	*v.*	促使
lord it over		称霸
capitalisation	*n.*	市值
apiece	*ad.*	各自
listed company		上市公司
brand loyalty		品牌忠诚度
analytics	*n.*	分析方法
ideal customer profile (ICP)		理想客户特征

premium	*a.*	高价的
tailored ad		定向广告
double whammy		双重重击
Royal Dutch Shell		荷兰皇家壳牌
ExxonMobil		埃克森美孚
decarbonization	*n.*	脱碳
rear-guard action		后卫战
inaction	*n.*	无所作为
venture out		冒险外出
turn up one's noses at		对……嗤之以鼻
plastic	*n.*	信用卡
card reader		读卡器
world-beating	*a.*	世界一流的
rebound	*v.*	反弹
be wary of		提防;担心
ramp up		增加
keep businesses afloat		使……正常运转
takeaway	*n.*	外卖食品

5.5 China Today 今日中国

Part One: Warm-up Activities

In Part One, you will practice typing sentences, paragraphs and passages dealing with today's China. Firstly you are supposed to read aloud and to identify the new words listed in Section A, and then try to practice typing the sentences, paragraphs and passages in Section B, C, D respectively under the teacher's guidance. While you are typing, please mark out the time you spend on each section and compare your results with your classmates.

Section A Vocabulary work

HSR (high-speed rail)		高铁
vinegar	*n.*	醋
lunar month		阴历月份
dramatically	*ad.*	显著地;戏剧性地
blossoming	*a.*	开花的
in the midst of		在……之中

abject	*a.*	可怜的;卑微的
temperate	*a.*	温和的;温带的
distinctive	*a.*	截然不同的
habitant	*n.*	居民;居住者
continental monsoon climate		大陆季风气候
Siberia		西伯利亚
Mongolian Plateau		蒙古高原

Section B Sentence practice

1. China is now the second-largest economy in the world.
2. Xi Jinping leads China to a moderately prosperous society.
3. China has scored a "complete victory" in its fight against poverty.
4. Transportation developed rapidly in China over the past 15 years.
5. China's HSR accounts for two-thirds of the world's total high-speed railway networks.
6. The advent of high-speed rail in China has greatly reduced travel time.
7. Chinese government will ensure social stability in Hong Kong and Macao.
8. The Spring Festival, also called the Chinese New Year, is China's most important festival that falls on the first day of the first lunar month each year.
9. China provides assistance to a lot of members of the global community.
10. The Chinese have a common saying, "Seven things in the house: firewood, rice, oil, salt, soy sauce, vinegar and tea".
11. The main forms of Chinese pre-school education are nurseries and kindergartens.
12. The Mid-Autumn Festival falls on the 15th day of the eighth lunar month.
13. In the November 2020 Top-500 list, a total of 212 China-made supercomputers were included.
14. Chinese people are well-known for their hospitality.
15. The number of giant pandas is increased dramatically under the Chinese government's protection.

Section C Paragraph practice

1. We celebrated the 70th anniversary of the founding of the People's Republic of China. This occasion has inspired a strong sense of patriotism among all Chinese people, creating a powerful force that will bring great victories for socialism with Chinese characteristics in the new era.
2. Chinese people are well-known for their hospitality. They like to invite friends to their homes for dinner. Usually, they cook as many special dishes as they can, which is a

normal practice of Chinese people.

3. There is a big gap between the temperatures of North and South China in winter. In North China, the land is sometimes covered with snow and ice in winter. When people in Harbin, a city in northeast China, go on a visit to the ice lantern park in severe coldness, people in Guangzhou, a southern city, are enjoying a blossoming spring.

Section D Passage practice

➢ Passage 1

After getting to know today's China more and more, you will find out that it is possible to eliminate abject poverty, it is at least possible to exponentially increase GDP per capita, it is possible to build fantastic infrastructure, it is possible to lead in science and technology, or even that it is possible to include minorities in a harmonious society.

CPC has built today's China in over seven decades. From the revolution to the reform and opening-up, to the most recent COVID-19 epidemic, they have already demonstrated they have the vision, competence, and strength again and again.

➢ Passage 2

The majority part of China lies in the north temperate zone, characterized by a warm climate and distinctive season, a climate well suited for habitants.

Most of China has a marked continental monsoon climate characterized by variety. From September to April the following year, the dry and cold winter monsoons blow from Siberia and the Mongolian Plateau, resulting in cold and dry winters and great temperature difference between northern and southern China. From April to September, warm and humid summer monsoons blow from the sea in the east and south, resulting in overall high temperature and plentiful rainfall, and little temperature difference between northern and southern China.

Part Two: Audio-typing

Now you are going to listen to the recording. Do not refer to your textbook while you are listening. Then you are supposed to listen to each section sentence by sentence once again and type what you hear at the same time.

Listen to the audio clip 5-30 and type the following short and simple sentences.

1. Nearly 850 million people live and work in mainland cities.

clip 5-30

2. Confucianism places a heavy emphasis on Chinese daily life.
3. China is still in the process of urbanization and industrialization.
4. 23% of the land was covered by forest in China.
5. China's automotive industry bloomed since 1980.
6. The Chinese government officially recognizes 56 ethnic groups.
7. Chinese zodiac has great importance.
8. The very first auto-making joint venture in China was established in 1984.
9. China's foreign exchange reserves accounted for about one thirds of the worlds' total.
10. RMB serves as a currency of choice for cross-border payments.
11. China opened its first high speed rail line in 2008.
12. By the year of 2030，there will be an 8 plus 8 structure railways.
13. Beijing，Shanghai，Guangzhou，and Shenzhen are the top 4 cities in China.
14. Jacky Chan is the most popular Chinese kung fu star.
15. According to the statistics，China has over 3 million stadiums.

Section B

clip 5-31

Listen to the audio clip 5-31 and type the following longer and more difficult sentences.

1. Capital Beijing is said to be the first city to hold both summer and winter Olympic Games.
2. In 2008，it was the first time China won the most gold medals out of any other countries，winning 51 in total.
3. By 2018，Chinese athletes have won 3,458 international competitions and have set 1,332 world records.
4. China continues to invest more in building sports areas and developing the sports industry.
5. The number of people using fitness app has grown nearly 16 times in 6 years，reaching 165 million in 2019.
6. Mountainous land and very rough terrains make up about 67% of Chinese territory，basins and plains about 33%.
7. China had a population of about 541.67 million in 1949，now reaching 1.41 billion.
8. There are many local Chinese operas，and some enjoy great popularity. Among them，Beijing opera enjoys the greatest reputation.
9. The restructuring of the economy and resulting gains have contributed to a more than tenfold increase in GDP since 1978.
10. China's forex reserves skyrocketed when China joined the World Trade Organization

in 2001.

11. The forex reserve of China overtook Japan in the top spot in 2006, hitting the trillion-dollar mark for the first time.
12. New China reached the urban rate of 60%, 70 years after its founding in 1949, when the rate was just 10%.
13. China's new Five-Year Plan aims to achieve green development with peak carbon emissions by 2030 and carbon-neutrality by 2060.
14. Three-North Shelter Forest Program is now forming one of China's biggest strips of planted trees.
15. The life expectancy reached 77.3 in 2019, which is significantly higher than the world average of 72.

Section C

Listen to the audio clip 5-32 and type the following paragraphs sentence by sentence.

clip 5-32

1. In the early days of the founding of New China, the infant mortality rate was 200 deaths per thousand births. By 1969, that had fallen to 83. And by 2017, the infant mortality rate had dropped significantly to 6.8 deaths per thousand. That's 22 points lower than the world average, and five points lower than the average of upper middle-income countries. Maternal mortality rate has also plunged from 150 per 100 thousand at the early stage of the founding of New China to 80 in 1991, and 18 in 2018.
2. China's actual use of FDI (Foreign Direct Investment) has increased 60-fold since 1983 to 163 billion US dollars in 2020, surpassing the US to become the largest recipient of Foreign Direct Investment globally. The country has made great progress in optimizing its business environment and stimulating market vitality through improving its laws, regulations and practices affecting foreign companies.
3. China is now able to feed 20% of the world's population with only 9% of the world's arable land. But it's not been an easy journey. Agriculture production was extremely cut down before the founding of the People's Republic of China in 1949 due to years of war and natural disasters. Since 1979 China has intensified agriculture reforms, and the grain output in 2020 was nearly 6 times that of 1949.

Section D

clip 5-33

Listen to the audio clip 5-33 and type the following passage sentence by sentence.

A

Space Exploration in China

When the Western world was modernizing and reaching technological breakthroughs during the first two industrial revolutions, China, amid domestic social chaos, missed the opportunity to get ahead. When the People's Republic of China was founded in 1949, the country only had around 50,000 technological personnel.

Many Western countries had already gained much experience in advanced areas, such as nuclear weapons and space exploration.

But China began to catch up fast. In 1964, the country successfully detonated its first atomic bomb, making China the fifth nuclear-armed nation. And in 1970, China launched its first space satellite Dongfanghong I, 13 years after the Soviet Union's Sputnik I.

Space exploration has been fast expanding on multiple areas. In 2003, China became the third nation to send an astronaut into space, after Russia and the US.

In 2019, its Chang'e-4 mission made the world's first-ever soft landing on the far side of the moon, a milestone in humanity's lunar exploration. And its home-grown Mars probe Tianwen-1 is aiming to have an all-around study of the Red Planet's geography after being launched in July, 2020.

These significant scientific breakthroughs were due to the hard work of scientists and a great deal of investment. Since 2013, China has been the world's second largest investor in research and development spending, after the US.

clip 5-34

Listen to the audio clip 5-34 and type the following passage sentence by sentence.

B

Current Education Accomplishment in China

Over the past 100 years, China has made significant progress in the education sector. In 1921, the government spent only 1% of its fiscal expenditure on education. The situation began to improve significantly in 1995. In 2012, educational spending was 4% of GDP and has grown ever since. In 2019, public spending on education reached a record of over 4 trillion RMB.

And the immense efforts paid off. China has almost wiped out illiteracy over the past century. In 1921, the illiteracy was nearly 90%, but the number dropped to less than 4% in 2020.

The achievement is largely due to the successful popularization of nine-year compulsory education. In just over 20 years, China achieved universal coverage in nine-year compulsory education, a goal that took some western countries nearly a century to attain. In 2017, the net enrollment rate in primary schools reached almost 100%. The number of public colleges and universities in China surpassed 3,000 in 2020, compared with no more than 55 in 1921.

And more notably, some top Chinese universities now have high global rankings. According to QS, the world's largest international high education network, seven universities from Chinese mainland have made it to the global top 50.

This accomplishment has put China among middle and high income countries, in terms of education. Meanwhile, the number of teachers at all levels of education almost doubled from less than 9 million in 1978 to over 16 million in 2017.

Listen to the audio clip 5-35 and type the following passage sentence by sentence.

clip 5-35

 C

Industrial Highlights

China was the only country to have a presence in all the categories, whether low-end or high-end, of the United Nations industry classification system.

However, it wasn't always like this. It missed the industrial revolution in the 19th century, and agriculture dominated the economy. In 1933, modern manufacturing accounted for only 2.5% of GDP. And industrial development is limited by years of war.

When New China was founded in 1949, the country was only able to produce some basic consumer goods. The making of industrial products like tractors, cars and planes was out of the question.

The new government vowed to restore its industries through a special five-year plan starting from 1953, and its industrial output grew steadily as a result. A series of such plans followed after that year. In 1978, through the reform and opening-up, leaders looked to take the country's industrial development to the next level. And the results were immediate, as the industrial output increased 16-fold on that in 1952.

And that was just a flavor of China's potential. By 2010, China has surpassed the United States to become the world's largest manufacturing hub.

China's industrial growth can be seen in the expansion of some sectors like steel and cement. For example, the crude steel output in 1949 wasn't enough to even make a kitchen knife for each family. In 2019, it accounted for more than half of the world's total output.

clip 5-36

Listen to the audio clip 5-36 and type the following passage sentence by sentence.

➢ D

Speech at a Ceremony Marking the Centenary of the Communist Party of China (Excerpt)

July 1st, 2021

Today, the first of July, is a great and solemn day in the history of both the Communist Party of China (CPC) and the Chinese nation. We gather here to join all Party members and Chinese people of all ethnic groups around the country in celebrating the centenary of the Party, looking back on the glorious journey the Party has traveled over 100 years of struggle, and looking ahead to the bright prospects for the rejuvenation of the Chinese nation.

On this special occasion, it is my honor to declare on behalf of the Party and the people that through the continued efforts of the whole Party and the entire nation, we have realized the first centenary goal of building a moderately prosperous society in all respects. This means that we have brought about a historic resolution to the problem of absolute poverty in China, and we are now marching in confident strides toward the second centenary goal of building China into a great modern socialist country in all respects. This is a great and glorious accomplishment for the Chinese nation, for the Chinese people, and for the Communist Party of China!

Today, a hundred years on from its founding, the Communist Party of China is still in its prime, and remains as determined as ever to achieve lasting greatness for the Chinese nation. Looking back on the path we have travelled and forward to the journey that lies ahead, it is certain that with the firm leadership of the Party and the great unity of the Chinese people of all ethnic groups, we will achieve the goal of building a great modern socialist country in all respects and fulfill the Chinese Dream of national rejuvenation.

Notes

space exploration		太空探索;空间探索
nuclear	*a.*	核能的
detonate	*v.*	引爆;使爆炸
atomic bomb		原子弹
Soviet Union		苏联
Sputnik I		人造卫星一号
fiscal expenditure		财政支出

pay off		取得成功
illiteracy	*n.*	文盲
nine-year compulsory education		九年义务教育
notably	*ad.*	显著地；尤其
presence	*n.*	存在；出席
tractor	*n.*	拖拉机
vow to		发誓要……；许愿
manufacturing hub		工业生产中心
cement	*n.*	水泥
rejuvenation	*n.*	复兴；恢复
in prime		在全盛时期

参考文献

[1] 董金玲,郝景亚,郑凌霄.国际商务函电双语教程[M].北京:机械工业出版社,2011.
[2] 焦微玲.外贸英语函电——从基础到实践[M].北京:电子工业出版社,2013.
[3] 廖华英.中国文化概况[M].2版.北京:外语教学与研究出版社,2015.
[4] 廖瑛,周炜.实用外贸英语函电教程[M].2版.北京:对外经济贸易大学出版社,2016.
[5] 裴沁.商务函电[M].武汉:武汉理工大学出版社,2018.
[6] 中国国际贸易学会商务专业培训考试办公室.跨境电商英语教程[M].北京:中国商务出版社,2016.

与本书配套的二维码资源使用说明

本书部分课程及与纸质教材配套数字资源以二维码链接的形式呈现。利用手机微信扫码成功后提示微信登录，授权后进入注册页面，填写注册信息。按照提示输入手机号码，点击获取手机验证码，稍等片刻收到4位数的验证码短信，在提示位置输入验证码成功，再设置密码，选择相应专业，点击“立即注册”，注册成功。（若手机已经注册，则在“注册”页面底部选择“已有账号？立即注册”，进入“账号绑定”页面，直接输入手机号和密码登录。）接着提示输入学习码，需刮开教材封面防伪涂层，输入13位学习码（正版图书拥有的一次性使用学习码），输入正确后提示绑定成功，即可查看二维码数字资源。手机第一次登录查看资源成功以后，再次使用二维码资源时，只需在微信端扫码即可登录进入查看。